Atoms and Powers
An Essay on Newtonian Matter-Theory
and the Development of Chemistry

HARVARD MONOGRAPHS IN THE HISTORY OF SCIENCE

HARVARD MONOGRAPHS IN THE
HISTORY OF SCIENCE

Chinese Alchemy: Preliminary Studies by Nathan Sivin

Leonhard Rauwolf: Sixteenth-Century Physician, Botanist, and Traveler by Karl H. Dannenfeldt

Reflexes and Motor Integration: Sherrington's Concept of Integrative Action by Judith P. Swazey

Atoms and Powers: An Essay on Newtonian Matter-Theory and the Development of Chemistry by Arnold Thackray

Editorial Committee

I. Bernard Cohen (chairman)
Donald H. Fleming
Gerald Holton
Ernst Mayr
Everett I. Mendelsohn
John E. Murdoch

Atoms and Powers

An Essay on Newtonian Matter-Theory
and the Development of Chemistry

by Arnold Thackray

Harvard University Press, Cambridge, Massachusetts, 1970

QD
14
.T53

© Copyright 1970 by the President and Fellows of Harvard College
All rights reserved
Distributed in Great Britain by Oxford University Press, London
Library of Congress Catalog Card Number 72-99521
SBN 674-05257-9
Printed in the United States of America

To Three Marys

Apologia

To call physics and biology ugly sisters would be ungracious and unfair. Even so, few would deny the Cinderella-like status of the history of chemistry within the larger discipline of history of science. There are many reasons for this status. The most obvious depend on the way in which theoretical advances in biology and physics have recently and profoundly affected the philosophical outlook of Western man. The resulting conceptual challenges have led many able practitioners to consider afresh the historical development of these sciences. Modern chemistry, confident and successful within a seemingly unshakable "reductionist" tradition, knows no such fundamental problems. Considered solely on the level of ideas, the science has for many years lacked that sense of present crisis from which true history is born. This lack has had its inevitable reflection in the lowly place accorded to chemistry by intellectual historians.

It is these historians who have dominated recent Western writing on the development of science. But science faces other challenges besides the internal and external intellectual ones generated by relativistic and quantum physics or evolutionary and molecular biology. The deep and difficult questions of the ends to which science should be put; of the relationships between science and technology, and between science and warfare; of recruitment, organization and morale among the highly-

Apologia

trained élite of professional scientists; of deployment, and of institutional restraints on creativity and free communication; all these are part of the deeper crisis that science now faces and that historians must make their springboard. Here the history of chemistry has peculiar relevance. Of all the scientific disciplines, chemistry is the one that from time immemorial has been most intimately related to the institutions, the technology, the immediate physical needs, and the techniques of warfare of society. So too chemistry was the science that first created the deliberate professionalism of the graduate school, the industrial laboratory, and the research team. It is therefore a subject of rich and continuing historical significance.

Within the perspective of the whole history of chemistry, the developments of the eighteenth century have usually been seen as possessing a particular importance. Ever since Lavoisier appropriated the phrase to his own work, "the chemical revolution" has been thought fundamental to the emergence of the science as we know it. Perhaps the clearest expression of the received view of this episode may be found in Professor H. Butterfield's *Origins of Modern Science*. Its eleventh chapter is entitled "The Postponed Scientific Revolution in Chemistry." The title epitomizes an attitude. Concentrating on the seventeenth-century triumphs of mechanistic thinking, this attitude does little justice either to the far richer intellectual origins of later chemical ideas, or to the complex interplay of cultural, social, intellectual, and economic factors that lie behind any of our modern sciences.

To recast the whole history of eighteenth-century chemistry would indeed be a major undertaking. The task is far beyond my powers. Yet such a rewritten history remains a compelling goal. The period witnessed the crucial stages in the transformation of chemistry from a motley collection of unconnected, and often contradictory, clusters of knowledge held by disparate social groups, into a coherent, professionalized and autonomous science. It may therefore be worthwhile briefly to review some of the major ingredients necessary to any new history.

Apologia

Given the present interests of historians of science, and the present primitive state of chemical history, it is natural if not wholly desirable to begin with the intellectual aspects of the science. Of these, the Cartesian legacy is certainly one, as is the Stahlian (or perhaps one should say Paracelsian) intellectual tradition. Then too, the increasing vogue for natural history is an important explanatory factor. The growing empirical stress, manifest in the experimentalism of a Black, a Scheele, or a Lavoisier, must also be given due place, as must the developing Romanticism characteristic of many later-eighteenth-century thinkers. But among intellectual factors, pride of place would surely go to Newtonianism. And it is with Newtonianism, and its curiously ambiguous relationship with the development of chemistry, that this essay is primarily concerned.

This is not for a moment to suggest that any history of early modern chemistry can rest content with discussing only such contributing intellectual traditions. The tensions and problems generated out of the relationship between chemical science and chemical technology are of obvious and central importance throughout this period. Again, the more directly sociological study of group loyalties and institutional patterns is fundamental to any full understanding of the intellectual shifts occurring as the social background, absolute numbers, economic status, professional orientation and technical abilities of chemical practitioners were undergoing dramatic changes. I hope to turn in future studies to aspects of these changes. Their fundamental historical significance, and their immediate contemporary relevance, scarcely require statement. But here and now my central concern is with intellectual factors, particularly Newtonianism. Thus this monograph is, as it were, a piece of prehistory, a clearing of the ground for future and far different studies. Yet it is, I trust, not without its own utility.

My present contention is that we cannot rightly appreciate the hopes, ambitions, research activities and theoretical concerns of many of the eighteenth century's ablest practitioners, until we take seriously the depth of their Newtonian convic-

Apologia

tions. My arguments may not command assent. I hope they will at least stimulate that debate and critical interchange which the history of chemistry so urgently needs. I hope too that as we begin to discern more clearly the intellectual framework of early modern chemistry, so we may start to appreciate the need for, and wider significance of, an analysis that also embraces those economic, technological, and social forces which so powerfully molded this emerging science.

Arnold Thackray

Edgar Fahs Smith Memorial Collection
in the History of Chemistry

University of Pennsylvania
Philadelphia
December 1968

Acknowledgments

In a study that ranges over several countries and a considerable period of time, a certain lack of supporting detail is an unfortunate but necessary evil. It therefore gives me great pleasure to acknowledge the aid and stimulus I have derived from the writings and advice of those who are at once scholars, colleagues, and friends.

I am specially indebted to three people: to Professor I. Bernard Cohen of Harvard University for his early aid to a neophyte among Newtonian scholars, his continuing encouragement over a period of years, and his valuable editorial comment on the manuscript; to Dr. Mary Hesse of Cambridge University for her unfailing help, lucid criticism, and lively interest in the problems of chemical history; and to Dr. J. R. Ravetz of Leeds University for first opening my eyes to the rich fascination of the history of science, and for perpetual provocation.

That my treatment in many places is less superficial than it would otherwise have been, is due principally to the publications and patience of Mr. G. Buchdahl (Cambridge University), Dr. M. P. Crosland (Leeds University), Dr. Y. Elkana (The Hebrew University, Jerusalem), the late Dr. F. W. Gibbs, Dr. F. Greenaway (Science Museum, London), Professor H. Guerlac (Cornell University), Professors A. R. and M. B. Hall (Imperial College, London), Sir Harold Hartley, Dr. D. M.

Acknowledgments

Knight (Durham University), Mr. J. E. McGuire (Leeds University), Mr. J. B. Morrell (Bradford University), Professor R. E. Schofield (Case Western Reserve University), Dr. W. A. Smeaton (University College, London), Professor R. S. Westfall (Indiana University), and Dr. D. T. Whiteside (Cambridge University). To all of them, as to those many other friends and colleagues who have helped to clarify my thinking, I offer my very best thanks.

A different debt, but no less real, is the one I owe to the perceptive writings of a succession of prewar French scholars. To those acquainted with their work, my obligations to Pierre Duhem, Emile Meyerson, Pierre Brunet, and above all to Hélène Metzger will be immediately apparent. I have also drawn freely on the encyclopedic and invaluable compilations of the late Professor J. R. Partington.

Like all fellow laborers, I wish to express my gratitude to those many officials who have so willingly aided my researches. Particular thanks are due to the librarians of: the American Philosophical Society; the Bodleian Library; the British Museum; Cambridge University Library; Kings College, Cambridge; Trinity College, Cambridge; Edinburgh University Library; the Chemical Society, London; the Royal Institution, London; the Royal Society of London; the Library of the Society of Friends, London; the Wellcome Historical Medical Library, London; the Manchester Literary and Philosophical Society; Manchester University Library; the National Library of Scotland; Christ Church, Oxford; Manchester College, Oxford; the University of Pennsylvania Library; Widener Library, Harvard College.

In its final form this book owes much to the stimulus provided by colleagues and students during a delightful year at Harvard. On a more tangible level it is a pleasure to acknowledge how financial assistance from the American Council of Learned Societies and the National Science Foundation greatly aided the closing stages of the work. Last but by no means least

Acknowledgments

let me record my indebtedness to the Master, Fellows and Scholars of Churchill College, Cambridge, for so generously supporting my earlier researches, while also providing the light reliefs of politics and wine.

A.T.

A Note on Citation

A full bibliography will be found at the close of this essay. References to it are made by author, year of publication, catchphrase and, where appropriate, page number—as Hesse (1961), *Forces and Fields,* 186. In cases where the context makes plain the title or nature of the work cited, the catchphrase is omitted. The *Dictionary of National Biography* is referred to throughout as *DNB,* and Partington's *History of Chemistry* as *Partington* (1961–64).

Individuals discussed at any length are identified by their years of birth and death, and references to the most recent scholarly study. In almost every case, additional biographical and bibliographical information may be found in either or both the *DNB* and *Partington* (1961–64). Explicit reference to these occurs only when other studies are not available.

For primary sources in foreign languages, contemporary English translations have been used as far as possible. In almost every case the translation has been verified against the original. Where published translations are not available, I have made my own, striving for fidelity to meaning rather than literal correspondence. Punctuation has been silently modernized and corrected throughout, and some of the wider variations between past and present orthography removed. In no case has the word order been altered, or italics introduced, without the fact being noted. Old Style dates have been left unaltered, save for the denoting of the year—for example, 20 March 1726/27 appears as 20 March 1727.

Contents

Foreword by I. Bernard Cohen xxi

1. Introduction: The Hidden Key   1

2. Newton: or Chemistry, Forces, and the Structure of Matter   8

    2.1. The Importance of Newton   8
    2.2. The Problem of Editions and Chronology   11
    2.3. The Internal Structure of Matter   12
        2.3.1. The *Principia*   14
        2.3.2. The *Opticks*   18
        2.3.3. The *De Natura Acidorum*   24
    2.4. The Ether   26
    2.5. Interparticle Forces   32
    2.6. Summary and Prospect   39

3. The Immediate Impact of Newton's Ideas   43

    3.1. Introduction   43
    3.2. Gregory, Pitcairne, and the Beginnings of Discipleship   45
    3.3. Cheyne, Mead, and the Medical Tradition   49
    3.4. The Keills, Freind, and Leibnizian Warfare   52
        3.4.1. The "Nut-Shell" Theory   53
        3.4.2. The Theoretical Discussion of Forces   67
    3.5. Early Experimental Physical Chemistry   74

Contents

4. The Spread of Newtonianism: A Tradition Established   83
    4.1. Introduction   83
    4.2. Newtonianism in France   84
       4.2.1. Early Contacts   85
       4.2.2. Visits, Controversy, and Newtonian Chemistry   88
       4.2.3. Newtonianism Established   95
    4.3. Newtonianism in Holland   101
       4.3.1. 'sGravesande   101
       4.3.2. Musschenbroek   105
       4.3.3. Boerhaave   106
    4.4. Newtonianism Established in Chemistry   113
    4.5. Some General Remarks   121

5. Speculative Systems   124
    5.1. Introduction   124
    5.2. Robert Green and Expansive and Contractive Forces   126
    5.3. The Ether, Electricity, Magnetism, and Heat   134
       5.3.1. Bryan Robinson and the Medical Tradition   135
       5.3.2. Gowin Knight: Attractive and Repulsive Matter   141
       5.3.3. Cleghorn and Caloric   147
    5.4. R. J. Boscovich and the *Theoria Philosophiae Naturalis*   150
    5.5. Clairaut, Buffon, and the Law of Attraction   155

6. The Problem of the Elements   161
    6.1. Introduction   161
    6.2. Chemical Elements and Mechanical Philosophy   162
    6.3. Homberg, Stahl, and Chemical Tradition   165
    6.4. Particle Size and the "Nut-Shell" Theory   176
    6.5. Elements and Atoms in Dutch and British Chemistry   179
    6.6. The French Situation   192

7. Quantified Chemistry: The Newtonian Vision   199
    7.1. Introduction   199
    7.2. The Battle for Quantification   202
       7.2.1. Buffon and the French School   205
       7.2.2. Torbern Bergman   218
       7.2.3. Wenzel and Richter   221
       7.2.4. The British Effort   223
    7.3. The Fading Vision   230

Contents

8. **British Popular Newtonianism and the Birth of the *New System*** — 234
   - 8.1. The Popular Lecturing Tradition 234
   - 8.2. Number Games and the Conceptualization of Chemical Composition 238
   - 8.3. Matter-Theory and Theology 244
   - 8.4. John Dalton's Scientific Education 252
   - 8.5. The Origins of the Chemical Atomic Theory 256
   - 8.6. A New System of Chemical Philosophy? 269
     - 8.6.1. Chemical Units 270
     - 8.6.2. Chemical Mechanisms 275

9. **Conclusion** — 279

   Bibliographic Note 285

   Select Bibliography 287

   Index 319

Figures

I	Homogeneous *v.* heterogeneous matter	19
II	The possibility of transmutation	25
III	Pitcairne to Gregory in 1707	47
IV	Newton revises the *Opticks*	63
V	The internal structure of matter	64
VI	Geoffroy's affinity table	91
VII	Boscovich's force-curve	153
VIII	Musschenbroek on internal structure	181
IX	Cullen's "quantification" of chemical mechanism	227
X	Dalton's illustration of his 1801 "theory of mixed gases"	260
XI	Dalton's view of chemical combination	265

following page 266

XII	Dalton's wooden atomic models
XIII	Modern atomic models

Table

1	Newton's works on natural philosophy	12

Foreword

The series of Harvard Monographs in the History of Science has been planned so as to bring together works that illustrate some of the main aspects of the development of science. Historians are generally aware that one of the most fruitful kinds of inquiries in this discipline in recent decades has been the study of the development of scientific ideas, in particular the way in which the precepts, the concepts, and the example of a single scientific figure may influence the actual growth of a scientific discipline. Many scholars have focused their attention on the life and work of Isaac Newton, a natural consequence of the exalted position attributed to him by his contemporaries, his immediate successors, and the historical critics writing over a period of two centuries. So many scholarly studies on Newton are currently appearing, in fact, that it is difficult to keep up with them; the whole operation has become likened to an international "Newtonian industry."

Until now, however, one topic of obvious importance has not been fully explored: the influence of Newton on the development of chemistry and of theories of matter. That this should be a major subject for research was suggested more than half a century ago by Pierre Duhem, and it was brought more forcibly to our attention by the penetrating and illuminating investigations of the late Hélène Metzger. In the decades since World

Foreword

War II, a number of scholars have made significant inquiries into certain aspects of this general topic, including—among others—A. Rupert Hall and Marie Boas Hall, J. E. McGuire, Henry Guerlac, R. E. Schofield, and myself. However, hitherto there has not existed a single full-scale volume exploring this topic.

Anyone who reads the works of John Dalton carefully is led at once back to the ideas of Newton, as well as to the actual developments in chemical science. In this framework Arnold Thackray has developed his monograph, beginning with the Newtonian concepts related to what we would call theory of matter, chemistry, and forces between particles; all of these together determine the varieties, the forms and conditions of matter, and the interactions between varieties of matter. He concludes with an examination of the conceptual framework of Dalton's chemistry, discussing the manner in which Newtonian ideas, in their full flowering, entered and influenced the main stream of our modern chemical science. Those who have read Thackray's published articles on Newtonianism and chemical science and his studies on John Dalton (notable for their wealth of new material and insights) will expect that his present monograph will be more than a chronology of events. They will not be disappointed. Here is presented to us a new aspect (new in the sense of never before having been fully explored) of the development of science in the Age of Enlightenment.

Historians of science who are familiar with the writings of Emile Meyerson are aware that studies of the history of chemistry may have great significance for our intellectual understanding of science and for our philosophical comprehension of the way in which scientific ideas grow. Curiously enough, despite the writings of Pierre Duhem and Hélène Metzger and Emile Meyerson, scholars concerned with conceptual analysis in the history of science, or the history of scientific ideas and concepts, do not usually turn to chemistry, save possibly in connection with the work of a few isolated figures such as Paracelsus (and his disciples), Boyle, Priestley, and Lavoisier, and an

Foreword

occasional topic such as the concept of element, the phlogiston and oxidation theories, or the concept of a vital force. Arnold Thackray's monograph is a timely reminder that chemistry and theories of matter should have a prominent place among histories of conceptual developments in science, along with work on the development of mathematics, astronomy, mechanics, general physics, and biology.

Like all good monographs, this one sets forth many fascinating questions for historians and philosophers of science to ponder, in addition to its main task of providing general information and a new insight into science during one of its most interesting periods. Not the least of the questions set before us here is that of determining whether the influence of a great man is always beneficial. Are certain failures to achieve universal systems due to grandiose schemes incapable of fulfillment (at least at the time in question, or under the conditions set), or to the fact that the influence of a man such as Isaac Newton may lead scientists along unfruitful paths, or cause them to ask unrewarding questions? Or do such failures arise simply because the disciples of an Isaac Newton may, in this case, have been poorer chemists than men working under other influences in other countries? Thackray discerns many ways in which the particular social and political or general philosophical background of men associated with Newton (and the Royal Society) may have been related—through the kinds of people they were—to the type of science these men produced. He thus raises significant questions concerning the relations between social history and the intellectual component of the development of science.

The reader will be grateful to Arnold Thackray for drawing attention to such pertinent matters as these. By providing us with new information and a new relevance for older knowledge, the present work may be said to fulfill the highest requirements of a monograph in the History of Science.

January 1970   I. Bernard Cohen

Atoms and Powers

An Essay on Newtonian Matter-theory
and the Development of Chemistry

Sir Isaac Newton alone has thrown more light upon the leading parts of natural philosophy & has done more, to the establishing a rational chymical theory, than ever was done before.

Joseph Black in 1785

1
Introduction:
The Hidden Key

... the key of the most hidden phenomena of chemistry,
and consequently of all natural philosophy.
P. J. Macquer in 1766

... a key to unlock the innermost sanctuaries of
Nature ...
Torbern Bergman in 1785

In stressing the fundamental importance of Newtonian ideas to chemical understanding, such eminent chemists of the Enlightenment as Macquer and Bergman were at once displaying and encouraging a widely held conviction. The conviction was that Newton's views of matter and methodology between them afforded the means by which chemical inquiry could be made a rational and fully predictive part of physical science. The very metaphor of these ideas as the "key" to chemical knowledge was in common use for the greater part of the eighteenth century.[1] Isaac Newton's own reference to "the more secret and noble works of Nature" invited the analogy. Reflecting the growing interplay between chemical theory and burgeoning manufactures, Guyton de Morveau was even to see Newtonian ideas on matter and its properties as "the key to all the operations that Art demands of Nature."[2]

Today the importance of these ideas within the history of chemistry is little recognized. The key once so eagerly employed is hidden behind historiographic monuments erected, as it were, at the height of chemical empire. It says much about the

---

[1] The earliest use of the metaphor appears to be in Cheyne (1702), *Theory of Medicine*, 24: "That stupendiously great man, Mr. Newton ... to him we owe the only key, whereby the secrets of Nature are unlock'd."

[2] Guyton (1777–78), *Elémens de Chymie*, II, vi; for Newton's usage, see Newton (1704), *Opticks*, Bk. II, 65.

Atoms and Powers

neglected state of this branch of the history of science that the perspective generated in the heroic age of chemistry—the second and third quarters of the nineteenth century—should still pass so largely unchallenged.

Though Newtonian ideas enjoyed a deep and lasting hold on chemistry, the fruits of the resulting research enterprise were few and unrewarding. Hence those nineteenth-century chemists who sought to write the history of their new, flourishing, autonomous, and highly successful science had little time for Newton and his influence. Not for them the hopes, the dreams, the arduous and unsuccessful experimental labors of a Guyton or a Kirwan. Instead their attention was focused on "positive science." The beginnings of such science were to be found not in speculative inquiries into internal structure and elective affinities, but rather in the antiphlogistic labors of Antoine Lavoisier and the chemical atomism of John Dalton. Generations of qualified historians, from Thomas Thomson and Hermann Kopp to J. R. Partington, thus found little worthy of note in Newtonian chemistry.[3] The whole enterprise, in the sense that earlier chemists had known it, quickly slipped from the historical record. One immediate object of this present work is to restore it to that record.

Thanks to recent research, we are now beginning to understand afresh how comprehensive was Newton's own view of his methodology, experiments, and theoretical constructs. He saw them as providing not only the essential frame for such macroscopic achievements as predictive astronomy and planetary theory, but also as the key to the microscopic world. Yet both in his own work, and in that of his followers, this key was a "hidden" one. Not only was the internal structure of matter by its

---

[3] Thomson's *History of Chemistry* mentions Newton only to explain why mathematics was the preferred pursuit of able men in early-eighteenth-century Britain. Hoefer does not refer to Newton at all, while Kopp's four volumes spare one paragraph for his chemistry. Von Meyer allows Newton two lines. Faithful to its nineteenth-century heritage, Partington's recent compilation gives only 10 pages to Newton, in contrast to 132 on Lavoisier and 73 on Dalton: see Thomson (1830–31), I, 303; Kopp (1843–47), II, 309; Hoefer (1842–43); von Meyer (1891), 132; *Partington* (1961–64), II, 468–477.

Introduction: The Hidden Key

nature inaccessible to direct observation, the empirical methods of approach were also hidden—in the sense of being uncertain, faltering, given to ambiguous results. Even so, it requires but a brief encounter with the writings of the period to realize the deep emotional hold that Newton's brilliant optical investigations and his dazzling quantification of the macrocosm together exercised on the eighteenth-century scientific imagination.

A recovery of the dominating influence of Newtonianism in all its forms will put many important aspects of chemical history into truer perspective. The present writer has elsewhere pointed to the enduring strength of the Newtonian tradition, to "the Newtonian dream" of quantifying chemical affinities, to the pervasive influence of the "nut-shell" theory of matter, and to the sharp political overtones of early Newtonian apologetic writing.[4] Professor R. E. Schofield has shown what different hues Joseph Priestley's "casual experimentation" assumes when seen in Newtonian light.[5] And the range of possibilities for further inquiry is of embarrassing scope, complexity, and significance.

In this essay, particular attention will be paid to the interrelation of chemical theory and Newtonian ideas in Britain. Such a choice of emphasis can be simply explained. It was within a peculiarly British context that Newton elaborated, and a century later Dalton transformed, those concepts of matter and its properties that were so strongly to influence research and debate on chemical units and mechanisms right down to the close of the nineteenth century. "Newton to Dalton" thus makes a natural period for study. Period and theme together demand a primary emphasis on British activity. However this emphasis should in no way be construed as denigrating (*inter alia*) the importance of Stahlian thought or the significance to the broader history of chemistry of that whole area of analysis,

---

[4] Thackray (1966d), *Newtonian Tradition and Chemistry;* (1968a), "Quantified chemistry"; (1968b), "Matter in a nut-shell"; (in press), "The business of experimental philosophy."

[5] Schofield (1966), "Priestley—philosopher," and (1967), *Scientific Autobiography.*

controversy, experiment, and propaganda that centers on Lavoisier.

Despite this choice of focus, the discussion that follows will not be confined to one country. Thought does not respect national frontiers. Yet scientific ideas are far from stateless citizens. They undergo subtle transformations as they enter, or fail to gain admittance to, foreign climes. It will therefore be necessary to pay attention to developments in France and Holland especially, insofar as these were affected by, and reacted on, the course of debate in Britain. In the particular case of Lavoisier, a more extended treatment would obviously have been rewarding. The present author has not attempted this treatment. And, lest the critic protest, it may be worth emphasizing the reasons for this omission.

One minor reason is that Lavoisier is today the subject of major, and as yet largely unpublished, scholarly studies.[6] This makes any sustained re-evaluation from outside unusually hazardous. More seriously, the brilliant experimental work and conceptual reformulations associated with his name were not directly aimed at either denying or replacing the Newtonian categories of thought, views of matter, or research priorities so widespread among his contemporaries. Hence his work is less than central to a study of Newtonian matter-theory and the development of chemistry. The task Lavoisier set himself was to reorganize the superstructure of the science rather than to deny its very Newtonian frame. That the impact of his work was far different is, as we shall see, one of history's ironies. It is also one reason why Lavoisier demands inclusion, if not extended treatment, in this present essay. Last of all, there is a more subtle motive. Not until we have paid attention to areas remote from both Lavoisier and phlogiston, can we hope to essay a meaningful alternative to that positivistic and mech-

---

[6] Most notably those by Professor Guerlac and his students. Guerlac (1961a), *The Crucial Year*, provides a foretaste of the larger work to come, while Gough (1968), "Lavoisier—new evidence," is the most recent contribution to specialist preliminary studies. Lavoisier's correspondence is also slowly being published: see Fric (1955–).

Introduction: The Hidden Key

anistic historiography which has so remorselessly accreted round these twin poles. It is just such a meaningful alternative that the present essay seeks above all to aid.

One final word of caution is in order. It concerns the term "Newtonian." For the eighteenth century, to join in paying homage to its master was one thing, to agree on the correct style for a life of discipleship, quite another. Newtonianism rapidly became many things to many men. Under one aspect, Newtonianism (or "the Newtonian philosophy") was frequently viewed as a matter of methodology. The avoidance of "hypotheses," the use of the method of analysis and synthesis, the refusal to avow causal mechanisms, the insistence on the mathematization of Nature—all these at various times and to varying extents were taken to be the essence of the Newtonian philosophy.[7] The present essay is not overtly concerned with these aspects of Newtonianism. The stress is rather on the views of matter and its properties felt to be sanctioned by, or reconcilable with, the works of the master.

In the senses in which it is used here, and in relation to chemistry, Newtonianism thus always implied two things. One was belief in the "inertial homogeneity" of matter, and consequently in a highly porous and internally ordered structure to all known chemicals (a belief which in extreme form became the "nut-shell" view of matter). The other was belief that for chemistry, as for astronomy, the route to an ordered, quantified, predictive science lay through the measurement of forces: in this case the forces of chemical affinity. It was this latter aspect that such men as Freind, Buffon, Bergman, and Kirwan had particularly in mind when they stressed how fundamental was the key to chemical science which lay in Newtonian theory.

The Enlightenment's vision of a fully predictive Newtonian chemistry, based on mathematical laws and experimentally measured forces, was not to be fulfilled. Instead the line of development, and successful quantification, lay with a theory of matter

---

[7] For a discussion of the variety of meanings read into "the Newtonian philosophy," see Cohen (1956), *Franklin and Newton*, 179–182.

outside the mainstream of Newtonian thought. Whatever his lip service to Newton, and whatever the complex roots of his ideas, John Dalton was far removed from the sophisticated theoretical discussions and experimental procedures common to the Newtonian chemists of his time. Instead his earliest world was that of the popular scientific lecturer. Utility, Quaker theology, and Scottish common-sense philosophy were his particular inheritance. This inheritance enabled him to undertake that refashioning of concepts, that denial of Newtonian tradition essential to any fruitful *New System of Chemical Philosophy*.[8] And, in combination with Lavoisier's natural-history-inspired administrative systematics, it was Dalton's autonomous, if philosophically unsatisfactory, models that were to underpin the triumphant nineteenth-century career of chemical science. Later chemists thus delighted to look back and laud Dalton as pioneer alongside, and almost equal to, Lavoisier. Only as we read the eulogies they composed, can we understand how important to them was the *New System*. Through it nineteenth-century chemistry enjoyed spectacular successes, together with an independence from reductionist models and categories unknown since the days of the Paracelsians. The work of Dalton thus makes a fitting terminal point for this present inquiry.

An understanding of the autonomous status of the "new" chemical science of the nineteenth century inevitably throws doubt on the relevance of schemes that would see the progress of chemistry primarily in terms of the triumph of physicalist or reductionist thinking. The profound failure of the Newtonian program forces us to seek far richer patterns of historical explanation. Not only does Dalton's own work reveal the power inherent in considerations peculiar to the continuing British conjunction of theology and natural philosophy. The whole of eighteenth-century chemistry, and especially its most successful aspects, must also be seen in relation to a greater diversity of

---

[8] The deliberately chosen title of his never-completed classic: see Dalton (1808–27).

Introduction: The Hidden Key

intellectual traditions.[9] Again, the question must be raised of the role in intellectual change of social, economic, and technological factors: a question which will be returned to in postscript. Here it must suffice to end this beginning by remarking that the continuing nineteenth-century undercurrents of skepticism about chemical atoms can only be fully understood when seen against that earlier background it is our present purpose to explore.

[9] A start—still largely unpursued—was made by Hélène Metzger's classic study: Metzger (1930), *Newton, Stahl, Boerhaave*. This in its turn built on the remarks in Duhem (1902), *Le Mixte*. A sketch that reveals something of the problems and possibilities is now available in Crosland (1963), "Eighteenth-century chemistry."

2
Newton: Or Chemistry, Forces, and the Structure of Matter

> But if anyone shall still retain a doubt of the worth and abilities of chymistry, to reward those who cultivate it: let him consider the practice and procedure of the happiest philosopher the world ever yet cou'd boast, the great Sir *Isaac Newton*: who, when he demonstrates the laws, the actions, and the powers of bodies, from a consideration of their effects, always produces chymical experiments for his vouchers; and when, to solve other phenomena, he makes use of these powers, his refuge is to chymistry.
> *Herman Boerhaave in 1718*

2.1. The Importance of Newton

For over a decade Isaac Newton [1] has been the central preoccupation of Western historians of science. A fresh understanding of his scientific achievement has become a major goal of intellectual history. The task has grown in importance with the realization that the treasure of unexploited manuscript is enormously rich; that the Presbyterian or positivist hagiography of a Brewster or a Mach is far removed from reality; and that Newton's science was enmeshed in a complex of fundamental philosophical issues. Much exciting work has been done. Already available, or in progress, are major re-evaluations of

---

[1] (1642–1727). Published just as the present study was completed, Manuel (1968), *Portrait of Newton*, is an iconoclastic biography which draws on and provides a convenient means of access to the researches of the past decade. The essays by a number of Newton scholars that are assembled in Texas (1968), *Annus Mirabilis Tricentennial*, cover most aspects of recent inquiry. The exponentially expanding literature on Newton defies reading, though Cohen (1960), Whiteside (1962), and Westfall (1966) offer useful bibliographical guidance to the intrepid. Brewster (1855), *Memoirs of Newton*, remains the standard account of Newton's life, while Mach (1893), *Science of Mechanics*, enshrines the positivist view of Newton and his achievements.

Chemistry, Forces, and the Structure of Matter

Newton's scientific background, mathematical development, optical work, and historical theories.[2]

Despite such impressive gains, we are far from a complete understanding of either Newton's immediate intellectual context or the chronological development of his scientific thought. We are even farther from relating his theology and alchemy to his more acceptable and "scientific" pursuits.[3] It might therefore seem premature to venture any exploration of such vexed subjects as Newtonianism and its influence on the course of natural philosophy. Yet the need for fresh thought on the development of chemistry neither can nor should await that millenial day when Newton scholars rest, content in an agreed new synthesis. And, as Herman Boerhaave's pronouncement so broadly hints, it would be folly to treat those developments apart from the heritage from Newton.

Newton's contributions to such fields as optics, astronomy, and mechanics are universally acknowledged. The profound effects that his ideas and achievements had on the methods and research programs of other sciences are less widely recognized. Quite apart from its methodological impact, his work was responsible for the central importance in eighteenth-century chemistry of three concepts. These are the concepts of a complex, hierarchical internal structure to matter, of imponderable fluids, and of quantifiable short-range forces. None of these ideas—except possibly the last—was original to Newton, but it was his development and use which rendered them so powerful and pervasive in later chemical thought.

When we seek to understand these ideas and their place in Newton's own work, we confront a tangle of difficulties. To begin with, Newton's thought may be studied on at least three levels. On one there are his writings as published and available to his contemporaries. On another there is his thought as modi-

---

[2] See respectively Westfall (1962), Whiteside (1967–), Sabra (1967), and Manuel (1963), for entrée to recent work in these areas.

[3] Some ways in which this might be done, and some insight into the new perspectives involved, may be gained from McGuire and Rattansi (1966), "The pipes of Pan."

fied in works published posthumously at various points in the following decades, and on the third that same thought as it may be reconstructed today from all these and the ten to fifteen million words of autograph manuscript which have survived the ravages of time.[4] Then again, Newton never wrote, and certainly never published, systematically and at length on what has been called his "theory of matter."[5] Yet it is increasingly clear that problems relating to the essential properties and micro-structure of matter were at the heart of his scientific concerns throughout his long, productive life. It is also clear that such problems remained central to eighteenth-century debate in its interrelated theological, philosophical, and chemical aspects: this the contributions of Berkeley, Boscovich, or Priestley immediately reveal.

Research into Newton's speculation on short-range forces, ethereal mechanisms, internal structure, and ultimate atoms is finally under way. Fresh information comes to hand almost daily.[6] Not enough is known yet for any large-scale reconstruction of his private thought to be possible. Happily, such reconstruction is largely irrelevant to our purpose. The course of later chemistry was guided and conditioned by those pronouncements of Newton that were published, or at least freely available, not by ideas undisclosed even to his closest acquaintance. Thus this study will be concerned primarily with Newtonian natural philosophers and their use, and abuse, of Newton's work, rather than with any attempt systematically to follow the development of his thought or to recapture the "real" Newton. Even so, some recourse will be made to the manuscripts for the

---

[4] There is no adequate bibliography of Newton, but see Gray (1907) and Babson (1950 and 1955). The checkered career of Newton's manuscripts is best discussed in the introduction to the first volume of Whiteside (1967–), *Newton's Mathematical Manuscripts*.

[5] Hall and Hall (1962), *Unpublished Newton*, III.

[6] For instance, the Wiles Lectures delivered by Professor Cohen in the spring of 1966 (to be published by the Cambridge University Press) dealt extensively with these themes. Other recent discussions include Cohen (1966), McGuire (1966 and 1967), and Thackray (1966d). Earlier forays into Newton's more directly alchemical and chemical writings include McKie (1942), Forbes (1949), Taylor (1956), and Hall and Hall (1958).

Chemistry, Forces, and the Structure of Matter

light they throw not only on the shifts in Newton's thought but also on his contemporaries' reception of his published work.

2.2. *The Problem of Editions and Chronology*

Newton's influence throughout the following century was as much in what others thought him to have said and established, as in what he actually wrote. That the period's major natural philosopher should be misunderstood is not in itself surprising. It is still less so when we realize the considerable textual alterations that characterize each fresh edition of both his major works.[7] These alterations reflect not only fresh evidence and maturing opinion but also unresolved tensions in Newton's thought. When to this we add the haphazard posthumous publication of a number of his minor writings, it is understandable how Newtonianism rapidly became a church of many denominations, each armed with appropriate citations from the written word.

The order in which Newton's works were published bears little relation to that of their writing. Much of his later thought became known before the earlier work from which it grew. To clarify the matter, Table I sets out, in order of publication, those of Newton's writings on natural philosophy that were available in print in the eighteenth century.[8] A glance at this table is sufficient to indicate the problems involved in discussing the impact of his ideas on contemporaries and successors.

The bases of Newtonian views of matter in Newton's own work will be elucidated in the sections that follow. Newton's thoughts on the fundamental structure of matter, interparticle forces, and the ether will each be considered in turn. Though the subjects are closely related, the relevant statements are sufficiently complex to demand separate treatment. The doctrines, hints, and speculations contained in the *Principia*, the *Opticks*,

[7] Drawn to the attention of modern scholars in Koyré (1955), "Des œuvres de Newton."

[8] Cohen (1958), *Newton's Papers and Letters*, provides facsimiles of the minor writings, with commentary.

Atoms and Powers

Table 1. Newton's Works on Natural Philosophy

Works	Published	Composed *
Papers on light and colors in the *Philosophical Transactions*	1672–76	1672–76
Principia, 1st edition (includes "Hypotheses" near start of Book III)	1687	1684–87
Opticks, 1st edition (includes first version of Queries 1–16)	1704	1675–1704
Opticks, Latin edition (includes first version of Queries 25–31) **	1706	1705
De Natura Acidorum	1710	1692
Principia, 2nd edition (In place of the "Hypotheses," Book III has the first version of the "Rules of reasoning in philosophy." Book III concludes with the "General Scholium.")	1713	1709–13
Opticks, 2nd edition (includes first version of Queries 17–24)	1717	–1717?
Opticks, 3rd edition (only minor changes from 2nd edition)	1721	–1721?
Principia, 3rd edition (includes final versions of "Rules of reasoning in philosophy," "General Scholium," etc.)	1726	1723–26
Lectiones Opticae	1729	1669–74
Opticks, 4th edition (only minor changes from 3rd edition)	1730	–1727?
The *Letter to Oldenburg*	1744	1676
The *Letter to Boyle*	1744	1679
The *Letters to Bentley*	1756	1693
Further papers on light and colors in Birch's *History of the Royal Society*	1756	1672–76

* These dates are tentative but sufficient for our purpose, which is to give an outline chronology rather than a definitive list.

** All Queries are referred to by their final rather than their original numbers. See section 2.3.3 below.

and the *De Natura Acidorum* will be central to the discussion. On the basis of the understanding thus provided, subsequent chapters will consider how inquirers into chemistry came to build on, puzzle over, misunderstand, and eventually turn away from the legacy Newton bequeathed to them.

2.3. The Internal Structure of Matter

The peculiar tensions in Newton's thought and work on the structure and properties of matter can be better appreciated

Chemistry, Forces, and the Structure of Matter

when we realize that he was not only the foremost natural philosopher of his day, but also "a man who belonged to the second generation of proponents of the mechanical philosophy." Receiving this philosophy as a birthright, he was willing to modify and transform it in a way its first expositors neither wished nor were able to do. The peculiarly British theological context of his early unease with the materialist connotations of both Descartes's theory of the plenum and Gassendi's atomism is now widely understood.[9] So is the important influence of Henry More on the way his thinking developed. Typically, a student notebook shows how in the mid-1660's, he "concluded in favor, not of Gassendi's atoms, but of Henry More's—a strange sort of particles as small as particles can be, *minima naturalia* or perfect parvitudes, as More called them, which attempt to combine the physical reality of the atom with the best features of the point."[10]

The endeavor to develop a form of atomism free from all taint of materialism and atheism was central to Newton's concerns. He was only partially successful, and never managed to codify or fully systematize his results. And the tensions still present in his thinking were to be magnified in the course of the clash with Leibniz. Still unduly influenced by Newton's image as a paragon of scientific virtue, we have perhaps failed to appreciate the weaknesses and ambiguities of the new ideas he advocated.

One of the great and continuing problems of his work was its failure to offer a philosophy of matter which could rival either of the popular mid-seventeenth-century mechanical philosophies in terms of simplicity of structure or analogical plausibility. The endeavor to create this new philosophy, and to reconcile God, forces, and brute matter, ran all through Newton's

---

[9] The most recent exploration is Guerlac (1963a), *Newton et Epicure*. Earlier discussions include Koyré (1957), *Closed World;* Metzger (1938), *Attraction Universelle;* Snow (1926), *Matter and Gravity;* and Burtt (1925), *Metaphysical Foundations*.

[10] Westfall (1962), "Newton's philosophy of Nature," 173–174. See also the extracts from More, and the discussion of "indiscerptibility," in Cohen and Koyré (1962), "Leibniz-Clarke correspondence," app. IV.

published work. First hinted at in his earliest papers on light and colors, the endeavor appeared in its most developed form in the final version of the Queries in *Opticks*. Discussion of the reflection, refraction, and transmission of light did after all raise questions about the mechanisms by which these effects were produced, and (in any "mechanical" explanation) the internal structure of matter on which transparency, translucency, and colors depended. Even so it was in the *Principia*, not the *Opticks*, that Newton chose to give his most extended formal treatment of the primary and secondary qualities of bodies, and the means by which one could reliably infer something about "the more secret and noble works of Nature."

2.3.1. *The Principia*

The reason that Newton chose to include in Book III of the *Principia*, in its 1713 form, a statement or "Rule of reasoning in philosophy" that centered on this "problem of transdiction" [11] lay in his desire to demonstrate the universal rule of the law of gravitation. This desire was natural in a work concerned with mathematical demonstrations applicable to the motions of the sun and planets. Yet, just because the "force" of gravity was only one of the various attractive and repulsive forces on which Newton sought to build his system, so the discussion could not fail to be of wider significance.

It has recently and ably been pointed out that the changes Newton introduced between 1687 and 1713 reflect in large measure his own struggle to define more closely the nature of transmutation. Whereas in the first edition of the *Principia* he was content to accept the prevailing opinion that "every body can be transformed into a body of any other kind, and can take on

---

[11] The phrase comes from Mandelbaum (1964), *Science and Sense Perception*, 61: Transdiction is "[using] data in such a way as not only to be able to move back and forth within experience, but to be able to say something meaningful and true about what lay beyond the boundaries of possible experiences." (And p. 66): "Any belief that ordinary material objects are actually composed of atoms, and the acknowledgement that these atoms are not capable of being perceived by our senses, commits one to a belief in transdiction."

Chemistry, Forces, and the Structure of Matter

successively all the intermediate grades of qualities," by 1713 he was more careful. The second edition stated only the related but distinct proposition that "the qualities of bodies, which admit neither intension nor remission of degrees, and which are found to belong to all bodies within the reach of our experiments, are to be esteemed the universal properties of all bodies whatsoever." [12]

This change in phrasing itself reflects the clarifying and firming of Newton's thought. More willing publicly to accept an atomistic philosophy in the 1690's than he was in the 1660's (for by this latter period he possessed that developed belief in forces as "God's guarantee" which earlier he lacked), Newton was naturally concerned to demonstrate that a world of flux was compatible with atoms of the sort he favored. His statement that some qualities were not susceptible of intension or remission marked the maturing of his thought on this. These invariant qualities were of course those by which matter or body was defined. As he elsewhere explained, they included the extension, shape, solidity, and inertia of the unchanging "primary particles."

Three things are immediately curious about this belief. All three were to haunt later Newtonianism.

The first curiosity is that, with his introduction of interparticle forces, difference of shape among the (unobservable) primary particles became at best a redundant hypothesis. Forces were capable of holding identical particles in a sufficient variety of patterns to explain all the manifold diversity of Nature. Yet so in thrall to "Moschus and Epicurus" was Newton, that it seems not to have occurred to him to abandon the gratuitous assumption of differing particle shapes.

Secondly, the inclusion of inertia among the primary properties of bodies was a radical step of profound significance. But

---

[12] Newton (1687), 402; (1713), 357–358; (1729), II, 203. The changes that this section of the *Principia* underwent, and the underlying shifts in Newton's thought, are complex to say the least. The matter is considered at length in Cohen (1966), "Newton's philosophy," and more especially in McGuire (1967), "Transmutation and immutability," which has been freely drawn on.

more than this, Newton's implicit and unyielding assumption that Nature possessed only one inertial constant (i.e., the assumption that *any* two units of *solid* matter if of the same volume were of the same inertial mass) was to have an enduring influence of the greatest importance. Such an assumption was purely arbitrary, as Roger Cotes soon pointed out.[13] It might agree admirably with prevailing beliefs about simplicity and harmony in Nature, but the existence of this inertial homogeneity was completely untestable. More than that, its acceptance automatically predicated belief in a vacuum, thanks to Newton's insistence on the proportionality of inertial and gravitational mass. Acceptance also permitted all those variations on a "nut-shell" theme that we shall meet in Chapter 3. As will become clear there, and in Chapter 6, this acceptance also militated against any viable concept of a chemical element.

The third curiosity about Newton's ideas was that he did not make plain, here or elsewhere, the status of the forces he at all times associated with matter. Apparently not primary qualities ("pray do not ascribe that notion to me" he replied when Richard Bentley was careless enough to assume gravity "essential and inherent" to matter), no more were they merely secondary qualities (for Newton nowhere allowed that it was possible to obtain matter without its associated "active principles"). Such a dilemma might be resolved with theological satisfaction by recourse to God—a fact which helps to explain the close association between dissenting academies and scientific curiosity so common in Hanoverian England—but it was unreasonable to suppose the whole European scientific community would be content with God. Hence the continuing dilemma, which we shall explore further in due course, about the number and nature of the fundamental forces, and the possibility of weightless fluids.

If we wish to examine more closely what the *Principia* has to say on the structure and properties of matter, it is natural to start with Definition 1. Its clear distinction between gravitational weight and inertial mass, and its stress on their proportionality,

[13] Edleston (1850), *Newton-Cotes Correspondence*, 65–69, 73, 75–76, 80, 210–211.

Chemistry, Forces, and the Structure of Matter

were fundamental contributions, regardless of how circular Newton's arguments might later appear. Yet, because of his assumption that all matter was inertially homogeneous, Newton's assertion of proportionality between weight and mass was to require further explanation and defense. This was provided in an important set of corollaries to Book III, Proposition 6. Like most of the early sections of Book III, these corollaries underwent substantial revision between the first and second editions, but their central message remained the same. Because this message was to become a dominating orthodoxy in Newtonian chemistry —an orthodoxy ineffectively challenged by Robert Green, and by the Hutchinsonian group—these passages require close scrutiny.

Newton believed experiments with pendulums established that all bodies whatsoever, when placed under identical conditions, exhibit weights that are directly proportional to their inertias (i.e., that gravitational mass is directly proportional to inertial mass). The inertial homogeneity of matter then allows the deduction of a direct and unchanging proportionality between gravitational mass and actual bulk of *solid* matter. Hence "the weights of bodies do not depend upon their forms and textures," as such weights can only vary with inertial mass, a property itself not subject to intension or remission. This conclusion leads to the rejection of what Newton came to call Aristotelian and Cartesian notions that ether might be altered into other matter by mere change of form.[14]

Having rejected the idea that weight might vary independently of inertial mass, Newton did develop an alternative explanation of why bodies of the same bulk exhibit different weights under the same conditions. This time the crucial passages, which clearly reveal Newton's own beliefs, are present only from the second edition onwards. Characteristically, the corollary most concerned is the only one of the set to be cast in conditional rather than affirmative form. It says that "if all the solid particles of all bodies are of the same density, nor can

[14] Newton (1687), 408, 410–411; (1713), 368.

be rarified without pores, a void space or vacuum must be granted. By bodies of the same density, I mean those, whose *vires inertiae* are in the proportion of their bulk" [15] (see Fig. I). This proposition, based on the inertial homogeneity of matter and the existence of gravity as a universal property, was entirely in accord with the position Newton had developed. Yet, presumably because it was not "inferred from the phenomena," it was left as a query rather than an assertion. Even so, there can be no doubt as to its importance in Newton's sight, or its role in the later general acceptance of an internal structure to matter.

2.3.2. The Opticks

To see further into the sort of internal structure Newton believed matter to possess, we must turn to the *Opticks*. Few eighteenth-century natural philosophers—and, for that matter, few later historians—were capable of holding the *Principia* and *Opticks* simultaneously in view. Yet though the two works possess very different styles, the difference between them lies not in the author, nor in the underlying beliefs, but in the subjects treated. The philosophy of matter which in the one is peripheral, in the other is a central and recurrent theme. Convinced as he was of the particulate nature of light and the dependence of color on particle size, in the *Opticks* Newton could not help but consider the nature of the particles he discussed. The brilliance of his experiments, already apparent in the 1670's but fully and systematically displayed only in the published book, went far toward insuring a sympathetic reception and wide diffusion for that view of the internal structure of matter with which these experiments were associated. In addition the Queries were themselves fascinating fruit of long research and speculation concerning matter and the forces which govern its changes and interactions.

It was in the Queries that Newton committed to print many of his deepest beliefs. The interrogative form was supposedly

---

[15] Newton (1713), 368. Newton (1726), 402–403 has a slightly variant reading. Translation quoted from Newton (1729), II, 224–225.

Chemistry, Forces, and the Structure of Matter

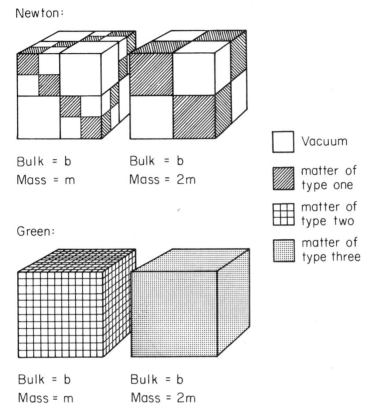

I. Homogeneous v. heterogeneous matter [for Green, see p. 129].

to make clear that they were speculations, not causes "inferred from the phenomena." But as Newton's prestige grew, so the Queries were increasingly taken as truth—a development their author seems to have appreciated. Already in the 1722 French translation (prepared with his active co-operation), it was stated that "although Sir Isaac Newton proposes his thoughts in the form of questions, penetrating eyes will not fail to see the solid foundations on which they rest." [16] And by 1734 J. T. Desaguliers (experimental assistant to Newton from 1713 to his death) could say that the *Opticks* "contain a vast fund of philosophy; which (tho' he [Newton] has modestly delivered under the name

---

[16] Coste (1722), sig. e iii [v,r].

of Queries, as if they were only conjectures) daily experiments and observations confirm." [17]

These statements give some indication of how the Queries' bold conjectures were to fascinate and intrigue later natural philosophers and condition their thinking and research. However, to study the Queries in isolation would be to divorce them from the background against which they must be understood. We shall instead approach them via the main body of the *Opticks*.

The opening words of Book I that "my design in this book is not to explain the properties of light by hypotheses, but to propose and prove them by reason and experiments," indicate the avowedly empirical nature of the work. Even such an ardent Cartesian as Fontenelle was to admit that "one advantage of this book, equal perhaps to that of the many new discoveries with which it abounds, is that it furnishes us with an excellent model of proceeding in experimental philosophy." In this Fontenelle reflected the general opinion. The opinion in turn reveals how great an influence the *Opticks* had on the development of experimental philosophy. And when we appreciate that Newton's work on light was bound in with a specific view of matter, we can begin to understand the importance of the main body of the *Opticks* to the development of chemical thought.

In Book II, Part III, Newton sought to apply to natural bodies the results of his work on the spectrum, and the colors of thin films. Consideration of the nature of reflection led to the conclusion that "between the parts of opake and coloured bodies are many spaces, either empty, or replenished with mediums of other densities." Newton thus formulated an argument for solid bodies containing many spaces between their particles, an argument also developed, on very different grounds, in the *Principia*. The dependence of color on particle size was next

---

[17] Desaguliers (1734–44), *Experimental Philosophy*, I, sig. cl^v. The reaction to the Queries in *Opticks* was first extensively discussed in Cohen (1956), *Franklin and Newton*.

Chemistry, Forces, and the Structure of Matter

formally stated and, continuing his inquiry, Newton established that "the parts of bodies on which their colours depend, are denser than the medium which pervades their interstices."

The way was then open for the key statement that "the bigness of the component parts of natural bodies may be conjectured by their colours." Several pages were occupied with just such conjectures, Newton ending with the observation that

> in these descriptions I have been the more particular, because it is not impossible but that microscopes may at length be improved to the discovery of the particles of bodies on which their colours depend . . . which if we shall at length attain to, I fear it will be the utmost improvement of this sense. For it seems impossible to see the more secret and noble works of Nature within the corpuscles by reason of their transparency.[18]

Nonetheless it was precisely these "more secret and noble works of Nature" that he went on to consider, developing and extending the type of model used by Boyle. The picture drawn was to fascinate and dominate chemists up to and even beyond the end of the eighteenth century.

Arguing from the carefully substantiated belief that actual physical contact between light rays and bodies resulted in the rays being "not reflected but stifled and lost in the bodies," Newton concluded that "bodies are much more rare and porous than is commonly believed," otherwise transparent bodies would not be possible. From this it was but a step to the statement that

> water is 19 times lighter, and by consequence 19 times rarer than gold, and gold is so rare as very readily and without the least opposition to transmit the magnetick effluvia, and easily to admit quick-silver into its pores, and to let water pass through it . . . From all which we may conclude, that gold has

[18] *Newton* (1704), Bk. II, 52 and 64–65.

more pores than solid parts, and by consequence that water has above forty times more pores than parts.[19]

Implicit in this statement was the belief that pure matter was inertially (and therefore gravitationally) homogeneous, or of a constant density. Only on this basis did it make sense to assume that because water is nineteen times lighter than gold, it is *by consequence* nineteen times rarer. This assumption of the unity and inertial homogeneity of matter was implicit in all Newton's work. His continuing acceptance of transmutation, and his desire to show how particles of widely different densities could be composed from uniform matter and void space must be understood in this context. Belief in this uniformity remained an unexamined presupposition of the Newtonian reductionist approach to chemical investigation right down the century.

In the 1704 edition of the *Opticks,* Newton was content to end his discussion of the porosity of solids by saying: "And he that shall find out an hypothesis, by which water may be so rare and yet not be capable of compression by force, may doubtless by the same hypothesis make gold and water, and all other bodies, as much rarer as he pleases, so that light may find a ready passage through transparent substances." [20] In 1706 two more pages were added to this discussion. Their purpose was to develop a model, showing how apparently solid matter might be built from complex hierarchies of pores and parts. Like so many of Newton's additions to later printings of his books, these passages were conjectural rather than affirmative in form. Their argument concerns the "nut-shell" view of matter, and will be dealt with in Chapter 3, in the context of early British Newtonianism and the bitter warfare with Leibniz.

In its final form the *Opticks* concluded with thirty-one Queries. The first sixteen of these, dealing with the properties of light, appeared in their earliest version in the original 1704 edition. Queries 25–31, which further elaborate these properties,

[19] *Ibid.,* 69.
[20] *Ibid.,* 69–70.

Chemistry, Forces, and the Structure of Matter

were added to the Latin edition of 1706 (numbered as 17–23), and Queries 17–24, which concern the ether, only appeared in 1717.[21] The 1717 revisions also included substantial alterations of and additions—especially of chemical discussion—to the previously published Queries. Query 31 in particular took on more the nature of a short chemical treatise (it occupied thirty-three pages) by the end of this, the last extensive revision.

It was in Query 31 that Newton chose to provide an extended and discursive account of his philosophy of matter. In this account he re-expressed the view that the smallest particles of matter cohere to compose bigger particles more weakly bound "and so on for divers successions, until the progression end[s] in the biggest particles on which the operations in chymistry, and the colours of natural bodies depend, and which by cohering compose bodies of a sensible magnitude." Only after detailed discussion of a host of experimental observations, did he make that well-known statement which begins: "All these things being consider'd, it seems probable to me, that God in the beginning form'd matter in solid, massy, hard, impenetrable, moveable particles, of such sizes and figures, and with such other properties, and in such proportion to space, as most conduced to the end for which he form'd them."

These particles of Newton, the ultimate particles discussed earlier in the same Query, were apparently true atoms, indivisible by any power in Nature. Made of uniform matter, they varied in size and shape, and therefore also in weight. They differed from the atoms of classical authors both in the hierarchical way in which observable particles were built from them, and in the forces associated with them—not only the inherent "vis inertiae," but also the active forces of gravity, fermentation, and cohesion at least.[22] An atomic philosophy with these two additions was to prove enormously appealing throughout the next century. It guided and molded the course of research. Yet

---

[21] Koyré (1960b), "Les Queries de l'Optique."
[22] Newton (1706), 337–338 and 343–345. Translation quoted from Newton (1718), 375.

just as the sort of "corpuscular philosophy" previously advocated by Boyle and Lemery had shown itself too simple a scheme to embrace the varied data of chemistry, so Newton's ideas were eventually to prove too complex for chemists to handle.

2.3.3. *The De Natura Acidorum*

To see the sort of chemistry Newton envisaged, it is only necessary to turn to the *De Natura Acidorum*. This short tract seems to have circulated in manuscript form for two decades before finally being published by John Harris in 1710.[23] It extended and formally stated some of the chemical notions implicit in the *Opticks*. Thus the observation that "all bodies have particles which do mutually attract one another: The summs of the least of which may be called particles of the *first composition,* and the collections or aggregates arising from the[m], primary summs; or the summs of these summs may be call'd particles of the *second composition,* &c," might contain nothing new except the labeling of the different orders of particles; but that was something.

Among a variety of speculations in the tract, one was that

> if the particles of the first, or perhaps of the second composition of gold could be separated; that metal might be made to become a fluid, or at least more soft. And if gold could be brought once to ferment and putrefie, it might be turn'd into any other body whatsoever. And so of tin, or any other bodies; as common nourishment is turn'd into the bodies of animals and vegetables.[24]

Just as in the Queries the transmutation of light into matter was considered and supported by a variety of analogies, so here the transmutation of gold was seen as a real possibility. This view was once again based on the fundamental unity of matter, yet

---

[23] See Turnbull (1959–61), *Newton Correspondence,* III, 205–214, for full details.
[24] Harris (1710), *Lexicon Technician,* introd. Facsimile reproduction in Cohen (1958), *Newton's Papers and Letters.*

Chemistry, Forces, and the Structure of Matter

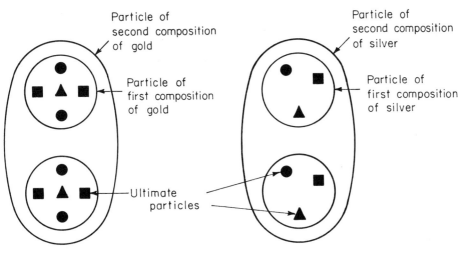

II. The possibility of transmutation.

the reason why Newton believed transmutation possible is more complex than first appears, and throws further light on his atomism.

For Newton, change of any sort was possible only by the separation, association, and motion of the ultimate permanent particles of matter. This view was perfectly in accord with classical atomism. These ultimate particles of different sizes and shapes were themselves immutable. Their shape, size, and inertial mass were not properties that could intend or remit. Hence the particles were not directly identifiable with any of the observed ingredients of the natural world, all of which were subject to change. Thus gold possessed its particular character not because it consisted of immutable gold atoms, as Dalton was later to hold, but because a variety of primary particles were combined into "particles of the first composition" of such a type and configuration as to create gold. Consequently gold, and all other observable bodies, were compound. If the particles of the first or possibly the second composition could be separated, they were of course transmutable (see Fig. II). Such transmutations might be observed in the irreversible changing of food into the bodies of plants and animals, and the action of acids on some earths. In contrast, normal chemical reaction proceeded on a more superficial level, affecting only the particles of the last composition. Hence gold, say, might be recovered unchanged from its solution in aqua regia.

Newton's atomism thus had a direct connection with chemistry, to which it gave a useful explanatory model. It both provided a mechanism for observed transmutations (irreversible reactions in modern terminology) and explained the difference between these and ordinary chemical reactions. A natural part of this viewpoint was the belief that all chemicals were in fact compounds, and that the ultimate particles or atoms were far below the level of direct observation. These beliefs were to remain unchallenged throughout the eighteenth century. Gowin Knight and R. J. Boscovich, realizing the redundancy of varied shape among unobservable atoms, developed schemes in which all the variety of chemical experience was derived from the interrelations of identical particles. But these new and simplified theories again accepted the inaccessibility of the fundamental particles, and the compound nature of all known chemicals.

2.4. *The Ether*

Newton's most fundamental contribution to the development of matter-theory was his replacement of the sort of "corpuscular philosophy" favored by, say, Boyle, with a view of Nature based on particles and the forces between them. In his more rigorously objective moments, he was careful to explain that the cause of the forces was exactly what he did not know. Yet he could not resist speculating on a possible "material" cause that would unify the observed diversity of forces.

That Newton had engaged in speculations on ether mechanisms from an early age is apparent from his manuscripts. Yet by the 1690's, he had moved so far from his earlier ideas as to refer somewhat contemptuously to "the vulgar supposition that gravity is mechanical." Such friends as Sir Christopher Wren were in turn reported as smiling "at Mr. Newton's belief that

Chemistry, Forces, and the Structure of Matter

[gravity] does not occur by mechanical means, but was introduced originally by the Creator." [25] By this time physical forces as the direct manifestations of God's presence and power in the world had become central to Newton's maturing scheme of theology, chronology, and natural philosophy. The deep roots and powerful fruits of this great preoccupation of his later days are only new beginning to receive a due appreciation. For our purpose the tone and implications of Newton's thought are beautifully and sufficiently caught in a note by his disciple David Gregory:

> 21 December, 1705. Sir Isaac Newton was with me and told me that he had put 7 pages of Addenda [in the 1706 *Optice*] . . . His doubt was whether he should put the last Quaere thus. *What the space that is empty of body is filled with?* The plain truth is, that he believes God to be omnipresent in the literal sense; and that as we are sensible of objects when their images are brought home within the brain, so God must be sensible of every thing, being intimately present with every thing: for he supposes that as God is present in space where there is no body, he is present in space where a body is also present. But if this way of proposing this his notion be too bold, he thinks of doing it thus. *What cause did the Ancients assign of gravity?* He believes that they reckoned God the cause of it, nothing else, that is no body being the cause; since every body is heavy.[26]

The *Optice* in its published form contained no such bold statements. However the stance and structure of its new Queries accurately reflected the position Gregory recorded.

Sometime after its publication and before 1713, Newton began to retreat (or perhaps we should say advance in new ways) from this more extreme position. The causes of the shift in his

---

[25] Scott (1967–), *Newton Correspondence*, IV, 267. Newton's remark is reproduced in Cohen (1966), "Newton's philosophy." For Newton's attitude toward the ether, see Guerlac (1967), "Newton's optical ether," and references cited there.

[26] Hiscock (1937), *Newton's Circle*, 29.

Atoms and Powers

thought are obscure. One suggestion is that they relate principally to the experimental work of Francis Hauksbee.[27] The rapidly escalating battle with Leibniz is another possible factor: the ether did after all serve to reduce an embarrassing variety of forces to effects of but one cause. Whatever the reasons, by 1713 Newton was again publicly referring to "a certain most subtle spirit which lies hid in all gross bodies." These enigmatic hints of the General Scholium to the fresh edition of the *Principia* were reinforced by the plain speaking of the new Queries in the second English edition of the *Opticks*. As the London *Newsletter* was to put it on 19 December 1717: "Sir Isaac Newton has advanced something new in the latest edition of his *Opticks* which has surprised his physical and theological disciples."[28]

It is of some interest to examine the properties of the ether as posited in 1717. The examination at once reveals that this ether did little to solve the central problem of Newton's whole system, that of how "matter might act where it is not." The assertion that "if the elastick force of this medium [the ether] be exceeding great, it may suffice to impel bodies from the denser parts of the medium towards the rarer, with all that power which we call gravity,"[29] was all very well. But the discussion did not specify how this elastic force was caused, or how the density gradients necessary to its continued utility were maintained, or why such an ether—composed of mutually repelling particles—was less of a conceptual worry than the mutually attracting particles its activity supposedly "explained." It is a mistake of later generations to believe that the ether was reintroduced primarily to provide a "contact-action" explanation of gravity. Its primary function was rather to simplify Newton's world system by referring the observed diversity of forces to the unity of one "certain most subtle spirit."

Newton was aware of the problems of infinite regress raised

---

[27] Guerlac (1963 and 1964).
[28] Quoted in Kargon (1966), *Atomism in England*, 138.
[29] Newton (1718), 326. The quotation is taken from the common 1718 "second issue" of the second edition of the *Opticks:* the rare "first issue" is dated 1717.

by an ether. As his manuscripts show, he was also very much alive to the threat to the unity of matter that any ethereal explanation of gravity necessarily implied. His private discussion of these things is worth quotation at length:

> If anyone should . . . admit some matter with no gravity by which the gravity of perceptible matter may be explained; it is necessary for him to assert two kinds of solid particles which cannot be transmuted into one another: the one [kind] of denser [particles] which are heavy [have gravity] in proportion to the quantity of matter, and out of which all matter with gravity and consequently the whole perceptible world is compounded, and the other [kind] of less dense particles which have to be the cause of the gravity of the denser ones but themselves have no gravity, lest their gravity might have to be explained by a third kind and that [again by a fourth and] so on to infinity. But these have to be very much less dense so as by their action to shake apart and mutually scatter the denser ones: by which means all bodies composed of the denser one would be quickly dissolved. And since the action of the less dense upon the denser will have been proportional to the surfaces of the denser, while gravity arises from that action and is in proportion to the matter of which the denser ones consist, it is necessary that the surfaces of the denser ones must be in proportion to their solid content, and therefore that all those particles must be equally dense and that they can neither be broken nor worn away nor in any manner destroyed: or else the ratio of the surfaces to the solid content, and consequently the ratio of gravity to the quantity of the matter would be changed. Therefore one must altogether determine that the denser particles cannot be changed into the less dense ones, and thereupon that there are two kinds of particles, and that these cannot pass into one another.[30]

[30] Quoted from a Newton manuscript of the early 1690's, reproduced in McGuire (1967), "Transmutation and immutability," 72–73.

Atoms and Powers

Despite these problems, in 1717 Newton did postulate just such "particles equally dense and that can neither be broken nor worn away nor in any manner destroyed," and as an automatic corollary, "two kinds of solid particles which cannot be transmuted into one another." In the light of the threat to the unity of matter, and the manifest failure of the ether to ease the problem of action at a distance, it is understandable how Newton's early followers (still battling entrenched Cartesianism and its wholly undesirable ether) tended to pass over the new 1717 Queries with embarrassed silence. Only when, in the 1730's and 1740's, they caught up with and passed beyond their master's own experimental inquiries into the electrical properties of matter, did they also begin to appreciate the explanatory utility of a repulsive-force imponderable-fluid ether. And as they appreciated this utility, so they turned back to Newton's earlier and far different speculations.

The *Letter to Boyle* of 1679 thus saw the light of day only in 1744. By then, no one worried that the letter spoke rather of an ether with variable properties, and one by no means clearly possessed of dynamic rather than kinematic qualities. It was sufficient to such busy systematizers as Bryan Robinson that the letter could be used to show how warmly Newton endorsed an ether. The even earlier *Letter to Oldenburg*, published at the same time, shows Newton's speculations in a prior and still more curious stage. This letter was actually written to correct some details of a 1675 paper, itself not published until 1756. It favored not one but several different ethereal spirits, and was quite happy to suppose that "the frame of Nature may be nothing but ether condensed by a fermental principle." So much for Newton's own later careful distinction between the ether and gross matter as "two kinds of solid particles which cannot be transmuted into one another!"

To confuse matters further, any mid-eighteenth-century natural philosopher zealous to know what Newton thought had also to cope with the four *Letters to Bentley*. These letters were written in 1692–93, in Newton's high theological period. They

Chemistry, Forces, and the Structure of Matter

were published in 1756 in a world concerned with electricity, imponderable fluids, and attraction and repulsion as essential properties of matter. Yet in the letters Newton denied that gravity was an essential and inherent property of matter. He also rejected as absurd the idea that matter could act at a distance, unless by "the mediation of something else, which is not material." As he enigmatically phrased it: "Gravity must be caused by an agent acting constantly according to certain laws: but whether this agent be material or immaterial, I have left to the consideration of my readers."[31]

In light of the changes in Newton's thought, the unresolved tensions, and the confusion caused by later publication of earlier ideas, it is not surprising that Newtonianism had developed a variety of conflicting and incompatible forms by mid-century. The major Newtonians were after all practicing natural philosophers, concerned to fit their current experimental and theoretical preoccupations into the frame provided by a brilliant set of insights, concepts, and methods, not historians concerned to recreate Newton's thought. Between the pious Priestley and the positivist Laplace there might seem to be little in common. Yet both, in meaningful senses, were Newtonians.[32] So also was Boscovich, with whom Priestley quarreled so bitterly over the theological implications of matter-theory, and so too was Buffon with his systematization that owed everything to a $1/r^2$ law, and nothing to imponderable fluids.

Though subject to all the ambiguities we have discussed, Newton's hints, speculations, and questions about the ether were greatly to influence the course of eighteenth-century thought, especially following the publication of the *Letter to Boyle*. Ethereal fluids which might or might not all be the same, but which were certainly exceedingly subtle and composed of mutually repelling particles, were to prove a fruitful source of

---

[31] Newton (1756), 31. Facsimile reproduction in Cohen (1958), *Newton's Papers and Letters*.
[32] But see Hahn (1967). "Laplace as Newtonian."

explanatory hypotheses. Throughout the second half of the century, these fluids played a vital part in the work of natural philosophers seeking to develop satisfactory theories to explain the complex phenomena of heat, light, fire, magnetism, and electricity.

2.5. Interparticle Forces

The name of Newton and the idea of a gravitational force are indissolubly linked. It is therefore surprising that so little is known about the origins and development of Newton's ideas of force. The roots of his differentiation between *active* principles (e.g., gravitation, fermentation, and cohesion) and *passive* ones (e.g., inertia) remain unexplored. His continuing unease about how many (and what) were the primary properties of bodies was well expressed in a 1690's note that "the essential properties of bodies are not yet fully known to us. Explain this by the cause of gravity, and by the metaphysical power of bodies to cause sensation, imagination, and memory, and mutually to be moved by our thoughts." [33] Newton's unease may have been partly hidden from his audience by the apparently confident and assured formulation of the 1713 "Rules of reasoning in philosophy." Yet that same unease was later published abroad in the *Letters to Bentley,* and the writings of such Newtonians as Buffon, Boscovich, and Bryan Robinson all reflect the continuing struggle to decide how many and which forces might be considered essential to matter.

His student notebooks reveal that Newton was early aware of such micro-scale problems as the cause of filtration, of air-expansion, and of "why the superficies of water is less divisible than 'tis within, insomuch that what will swim in its surface will sink in it." [34] The notebooks also reveal his dissatisfaction with the "matter and motion" explanations of orthodox me-

---

[33] Quoted from a manuscript reproduced in Cohen (1966), "Newton's philosophy," 181.

[34] Quoted in Westfall (1962), "Newton's philosophy of nature," 180.

Chemistry, Forces, and the Structure of Matter

chanical philosophy. The development of force explanations for these, and more directly chemical phenomena (apparently not yet among his concerns in 1664), was to become one of Newton's major preoccupations, and a driving force in his search for a natural philosophy more satisfactory than any then available. This preoccupation, this particular driving force, was every bit as central to Newton's concerns as were astronomical problems. That it resulted not in a *Principia* but in the fascinating display of experimental and observational virtuosity that filled and overflowed the 31st Query in *Opticks* is testimony rather to the magnitude of the problems involved than to their peripheral place in Newton's thought.

Already in the preface to the first edition of the *Principia,* Newton's maturing personal view of the mechanism of Nature could be seen in the statement that he was "induced by many reasons to suspect that they [the phenomena of Nature] may all depend upon certain forces by which the particles of bodies, by some causes hitherto unknown, are either mutually impelled towards each other and cohere in regular figures, or are repelled and recede from each other." [35] The *Principia* being a work devoted to mathematical demonstrations and discussion related to the force of gravity and its effects, these other "certain forces" were not prominent. Nonetheless they were mentioned.

Propositions 85 and 86 of Book I dealt with forces which decrease "in more than a duplicate ratio of distance," while Book II, proposition 23, considered particles whose centrifugal forces were "reciprocally proportional to the distances of their centers." The relationship between Boyle's law and this latter proposition bears more detailed examination. Newton's favorite simile for the ether was how it is "much like air in all respects, but far more subtle." Indeed his early acquaintance with "Mr. Boyles receiver" and the 1660 *Spring of the Air* no doubt help to explain his own insistence on the explanatory possibilities inherent in the ether's "elasticity." At what stage Newton came to view this elasticity as itself the product of a repulsive force

[35] Translation quoted from Newton (1729), I, sig. A 2^r.

Atoms and Powers

rather than of the coiled springs of orthodox mechanical philosophy, we do not know. What we do know is that by 1687 he thought it worthwhile to provide a mathematical proof that "particles flying each other with forces that are reciprocally proportional to the distances of their centers, compose an elastic fluid, whose density is as the compression."

Newton was careful to point out that "whether elastic fluids do really consist of particles so repelling each other, is a physical question. We have here demonstrated mathematically the properties of fluids consisting of particles of this kind, that hence philosophers may take occasion to discuss that question."[36] In view of the way his own thought had developed, of his use of ether-air analogies, and of his later resort to the air as an instance of the existence of repulsive forces, this statement seems disingenuous. But even if he were sincere in the disclaimer, the triumph of Newtonianism itself ensured that later readers mistook the question for the demonstration. From Hales through Lavoisier to Dalton, they all assumed Newton to have proved that gases do consist of mutually repelling particles in equilibrium.

In the early 1690's Newton contemplated a second edition of the *Principia*. Among its aims was a full revelation of the concordance between his system of atoms and gravity as a now definitely immaterial power and the belief of the Ancients. As David Gregory put it, "he will spread himself in exhibiting the the agreement of this philosophy with that of the Ancients and principally that of Thales. The philosophy of Epicurus and Lucretius is true and old, but was wrongly interpreted by the Ancients as atheism."[37]

The *Principia* was not the only benefactor from this attention, for at the same period Newton also worked on a version of the *Opticks*. This version consisted of four books, the fourth

---

[36] Newton (1687), 301, 303. Translation quoted from Newton (1729), II, 77–79. Newton's varying discussion of the magnetic force-law, in the successive editions of the *Principia*, is considered in section 3.5. below.

[37] Turnbull (1959–61), *Newton Correspondence*, III, 384.

Chemistry, Forces, and the Structure of Matter

having a conclusion of five hypotheses. The second of these was to express the essence of Newton's natural philosophy:

> And if Nature be most simple and fully consonant to her self she observes the same method in regulating the motions of smaller bodies which she doth in regulating those of the greater. This principle of Nature being very remote from the conceptions of philosophers I forbore to describe it in [the *Principia*] least I should be accounted an extravagant freak and so prejudice my readers against all those things which were the main design of the book: but and yet I hinted at it both in the preface and in the book it self where I speak of the inflection of light and of the elastick power of the air: but the design of that book being secured by the approbation of mathematicians, I had not scrupled to propose this principle in plain words. The truth of this hypothesis I assert not, because I cannot prove it, but I think it very probable because a great part of the phaenomena of Nature do easily flow from it which seem otherways inexplicable: such as are chymical solutions, precipitations, philtrations . . . volatizations, fixations, rarefactions, condensations, unions, separations, fermentations, the cohesion, texture, fluidity and porosity of bodies, the rarity and elasticity of air, the reflexions and refraction of light, the rarity of air in glass pipes and ascention of water therein, the permiscibility of some bodies and impermiscibility of others, the conception and lastingnesse of heat, the emission and extinction of light, the generation and destruction of air, the nature of fire and flame, the springinesse or elasticity of hard bodies.[38]

Thus Newton displayed his full conviction, in a way in which, for fear of being accounted an "extravagant freak," he never did in his published work. The nearest he came was of

---

[38] Quoted from a manuscript reproduced in Cohen (1966), "Newton's philosophy," 180.

course in the Queries that slowly piled up on the main bulk of the *Opticks,* till they almost obscured the work itself. These Queries were rich in purely chemical discussion—far richer than "Hypothesis 2" of the 1690's draft. Not unnaturally, they were to be of more direct interest to later chemists than almost any other part of Newton's published work. Though the earlier Queries did include some chemistry and some material on short-range forces, the quotations that follow are all drawn from Query 31—the longest and richest Query, and the one that constituted Newton's philosophical last will and testament to the learned world.

The Query opens with a passage that displays Newton's belief in the explanatory power of forces, his philosophy of Nature, and his public avoidance of "hypotheses." As Newtonianism became established in experimental philosophy, it was to this Query, and especially to its opening passage, that reference was increasingly made for validation, inspiration, and direction in the Newtonian research program.

> Have not the small particles of bodies certain powers, virtues or forces, by which they act at a distance, not only upon the rays of light for reflecting, refracting and inflecting them, but also upon one another for producing a great part of the phaenomena of Nature? For it's well known that bodies act upon one another by the attractions of gravity, magnetism and electricity; and these instances shew the tenor and course of Nature, and make it not improbable but that there may be more attractive powers than these. For Nature is very consonant and conformable to her self.[39]

Much of the rest of the Query was devoted to a discussion of these "certain powers, virtues or forces," especially in their chemical aspects. For instance, Newton asked how "when salt of tartar runs *per deliquium,* is not this done by an attraction

[39] First published in Newton (1706), 322. English version quoted from Newton (1718), 350–351.

Chemistry, Forces, and the Structure of Matter

between the particles of the salt of tartar and the particles of the water, which float in the air in the form of vapours?" Or again, "when salt of tartar *per deliquium,* being poured into the solution of any metal, precipitates the metal . . . does not this argue that the acid particles are attracted more strongly by the salt of tartar than by the metal, and by the stronger attraction go from the metal to the salt of tartar?" It was not only solution and precipitation that might be referred to variations in attractive force. All the phenomena of chemistry, from the explosion of gunpowder and the formation of crystals, to the differing effects of *aqua regia* and *aqua fortis* on silver and gold, could be elucidated by recourse to this force, as Newton rapidly indicated.[40]

It is little wonder that the 1706 publication of the Latin version of what was to become Query 31 was a signal for that outburst of chemical speculation of which John Freind's treatise remains the classic example. Nor is it surprising that Newton's attempt to list the metals in the order of their attractive powers was to fascinate E. F. Geoffroy and lead, through the latter's table, to the host of late-eighteenth-century attempts to quantify the force of chemical affinity. In the 1717 edition of the *Opticks* the powerful effect of these speculations on short-range attractive forces was further strengthened by the addition of yet more experimental evidence taken from the work of Francis Hauksbee. It was in this edition that the conclusion was first drawn that "there are therefore agents in Nature able to make the particles of bodies stick together by very strong attractions. And it is the business of experimental philosophy to find them out." [41] This same message had been implicit in 1706. Indeed it had been there right from 1687, for those with eyes to see.

Newton's speculations were not limited to attractive forces, as the following passage, from the latter part of Query 31, makes

---

[40] Newton (1706), 322–323, 325–327, 329, 334 etc. Cf. Newton (1718), 350–351, 353–355, 357–358, 363 etc.

[41] Newton (1718), 369.

plain: "Since metals dissolved in acids attract but a small quantity of the acid, their attractive force can reach but to a small distance from them. And as in algebra, where affirmative quantities vanish and cease, there negative ones begin; so in mechanicks, where attraction ceases, there a repulsive virtue ought to succeed." Examples of such a virtue were then produced, as in the reflection of light, the expansive power of air, and the fact that "flies walk upon the water without wetting their feet." "And thus," Newton could say, "Nature will be very conformable to her self and very simple, performing all the great motions of the heavenly bodies by the attraction of gravity which intercedes those bodies, and almost all the small ones of their particles by some other attractive and repelling powers which intercede the particles." [42]

To eighteenth-century experimental philosophers the implications of this statement were plain, and their work was correspondingly aimed at the completion of that inquiry into forces which Newton had begun. To understand, to quantify, to reduce the whole of Nature to simple laws of the form $1/r^n$ as Newton had reduced astronomy to a $1/r^2$ law: this was the compelling, the enticing, indeed the almost bewitching goal of so much work, especially in chemistry. In such work the *specific* attractions experimentally recorded between different chemicals were not of fundamental importance. The purely empirical chemist might, one supposes, delight in the mass of factual information reposing in the resultant "tables of affinity." But for most of the century the interest and challenge lay in the thoroughly Newtonian and reductionist task of uncovering the general mathematical laws which governed all chemical behavior. And in moments of doubt or despair, there was always the example of "the happiest philosopher the world ever yet cou'd boast" [43] to encourage chemists in the continuing labor.

---

[42] Newton (1706), 338 and 340–341; English version quoted from Newton (1718), 370–372.

[43] The judgment was Boerhaave's in 1718, as quoted in Shaw (1725), *Boyle's Works,* III, cclx.

Chemistry, Forces, and the Structure of Matter

2.6. Summary and Prospect

In the preceding sections, those ideas on the internal structure of matter, the ether, and interparticle forces, which were implied if not always fully developed in Newton's published works, have been outlined. Some of the difficulties raised by his speculations on an ether have been considered, and his philosophy of matter explored. Newton combined classical atomism and his own concept of gravity in his explanation of density-variation. His belief in a hierarchical internal structure to matter was not new, but the way it was developed and used gave it a new significance in the eyes of later philosophers. A particulate philosophy that ranged from light itself to those large particles on which colors depended, and that took in the complex phenomena of chemical reaction and transmutation, could not easily be ignored.

Significantly, though Newton's work could provide a basis for chemical explanations, his primitive particles bore no direct relation to any observable chemicals. They were able to assume whatever properties future investigators were to find convenient. Hence it is not surprising that eighty years later, after many and varied Newtonian speculations had been put forward, Lavoisier was to declare "that if, by the term elements, we mean to express those simple and indivisible atoms of which matter is composed, it is extremely probable we know nothing at all about them." [44] Later still it was to be Dalton who, by identifying the terms "atom" and "least particle of a chemical element," rejected this whole tradition and finally brought atoms to the level where their properties could be quantitatively inferred from chemical experiments.

When, just under a century after the publication of the *De Natura Acidorum,* the "chemical atomic theory" was to be announced to the world, the scandal and the novelty of it lay not in the idea that atoms of different weights existed and were at the basis of observed phenomena. The novelty lay rather in

---

[44] Lavoisier (1790), *Elements of Chemistry,* xxiv.

the suggestion that matter could exist in inertially heterogeneous forms, that the variety of these forms was exactly as the variety of known chemical elements or atoms, and that the relative weights of the atoms could be quite simply deduced from known weight relationships. For, as we shall see, Newton's ideas, themselves partly the inheritance of earlier thinking, still formed the basis of the early-nineteenth-century chemistry of such varied expositors as A. F. de Fourcroy and Humphry Davy.

It is difficult to decide how important to Newton himself were his speculations on the ether. Of their importance to later philosophers there can be no doubt. This importance was twofold.

Firstly, these speculations led to ideas on imponderable fluids which were to challenge the role of weight measurements in chemistry throughout the following century. In this sense phlogiston was not a theory to be overthrown, but a concept to be renamed caloric. Even Dalton was not to avoid or misinterpret this aspect of Newtonian tradition, but only to complete that shift towards chemistry as the science of what can be *weighed,* that Hales, Black, and Lavoisier had all assisted. The crucial difference was to be that with Dalton, for the first time, this macro-scale weight measurement was clearly allied to a purely *chemical* conceptual model—that of indivisible atoms of heterogeneous matter. To eighteenth-century Newtonian chemists, weights were important but far less fundamental than the short-range interparticle forces on which they believed chemistry really depended. Only with Dalton was weight measurement to become *the goal,* as well as *a means,* of chemical investigation.

Secondly, Newton's speculations on the ether were to help immensely in the organization of a vast and increasing mass of experimental data on heat, light, fire, magnetism, and electricity. In particular the Newtonian ether served as a model for the caloric theory of heat. This theory became established as part

Chemistry, Forces, and the Structure of Matter

of Lavoisier's new view of combustion, and subsequently proved essential to the development of Dalton's atomic theory. Paradoxically it was thus the ether, the least developed of Newton's ideas about matter, which alone became a part (however briefly) of the autonomous and antireductionist chemical science of the nineteenth century. In this essay it is not possible to discuss the experimental researches that were involved in these developments. The theoretical importance of the ether, and its relationship to ponderable matter, will, however, be considered in Chapter 5, in connection with speculative Newtonian systems.

Newton's other great contribution was the idea of interparticle forces. His ambiguity over how many, and which, of these forces were essential was to cause continuing confusion and debate. Even so, the concept of attractive and repulsive forces between the smallest particles of matter had an enormous effect on the course of chemistry. As we shall show, it is a mistake to suppose that chemists of the Enlightenment were indifferent to the quantification of their science. The case is rather that, completely dominated by the enormous success of Newtonian celestial mechanics, they valiantly strove to perform a similar quantification of chemical force-mechanisms. In contrast to this physicalist program, Dalton's method of quantification was intimately linked to his vision of independent chemical atoms. Whatever the philosophical problems posed by his theory of matter, Dalton's ideas were chemically successful. Hence they have enjoyed the homage of history, unlike the philosophically more coherent, if less successful, reductionist schemes of the Newtonians.

Before discussing the full-fledged tradition of the later part of the century, we must deal with the earlier development and diffusion of Newtonian natural philosophy. And first of all we must consider the impact of Newton's ideas on a developing British circle of friends, fellow philosophers, sycophants, and hangers-on. Their propagation of a more developed theory of

Atoms and Powers

the internal structure of matter—the "nut-shell" theory—demands detailed consideration, as do their theoretical and practical attempts to quantify chemical mechanisms.

From this discussion it will become abundantly clear that the failure to quantify chemistry in the eighteenth century was not owing to any lack of ideas on the possible relations between different particles, any inability to handle algebraic expressions, or any unawareness of the complex factors involved in such a process as solution. All this is obvious from the work of such early Newtonians as the Keills and John Freind. It seems that the difficulties lay not in the realm of pure ideas, but rather in the more subtle region of knowing which ideas were capable of exploitation, given the prevailing level of chemical knowledge and practice. It was to take another hundred years of argument and painstaking experiment before chemists were able to agree upon a set of elements that could be isolated in the laboratory. More important, it was not until the work of Berthollet had shown how unattainable was the Newtonian dream of understanding chemistry through the quantification of interparticle forces that chemists were ready to pursue alternative goals and alternative views of matter. But these are subjects for later chapters. Now we must turn to the diffusion and development of Newtonian ideas.

3
The Immediate Impact of Newton's Ideas

> For Gods sake keep Sir Isaac Newton at work, that we may have the chymical business, his thoughts about God, more of vacuum which he promised to me at Cambridge, that of hardness or greatest attraction, (the matter of atoms) and elasticitie if he pleases.
> *Pitcairne to Gregory in 1707*

3.1. Introduction

The wider philosophical implications of the *Principia* were still hidden in 1687, thanks largely to Newton's morbid fear of being thought an "extravagant freak." Even the mathematical demonstrations of the book were accessible only to the elect few. These few were quick to appreciate the profound significance of Newton's work. Typical of the substance if not the style of their reaction was David Gregory's note of 2 September 1687 about how "I think myself oblidged to give you my most hearty thanks for having been at the pains to teach the world that which I never expected any man should have knowne . . . and tho your book is of so transcendent fineness and use that few will understand it, yet this will not I hope hinder you from discovering more hereafter to those few." [1]

Of greater interest for our purposes is the group that began to form round Newton as his abilities became more widely known.[2] Ranging from his favorite, Nicolas Fatio de Duillier (later to be pilloried for theological fanaticism), through the

---

[1] Turnbull (1959–61), *Newton Correspondence*, II, 484.

[2] A study of this group is now available in chapter 13 of Manuel (1968), *Portrait of Newton*. The present author's work was independent of and completed before Manuel's study appeared, though the conclusions are compatible. Hiscock (1937), *Newton's Circle,* is an invaluable source for the study of the early Newtonian group.

smooth-tongued and flattering David Gregory to the irascible and intemperate William Whiston, the membership of this group was at best diverse and shifting, and its scientific ability widely varied. Nonetheless this British "circle" was of prime importance in the diffusion of Newtonian ideas, especially after the master's 1696 move to London increased the possibilities for fame, intrigue, and flattery. Again, Newton's 1703 assumption of the presidency of the Royal Society (a society he nursed conscientiously, administered zealously, and ruled firmly but without popularity) gave great scope both to his fertility in devising experiments for performance by the Society's demonstrator, and his genius for clandestine operations in the warfare with Leibniz. It is within this latter context that we can best understand the tone and content of some of his additions to the *Opticks* and *Principia,* and the supporting volleys of Newtonian orthodoxy loosed by, among others, John Keill, John Freind, and John Harris.

Much early Newtonian publication took place for an immediate and bitter polemical purpose. The authors and arguers were typically mathematicians, physicians, and popular experimental lecturers with a strong interest in natural philosophy in its new, prestigious, and unashamedly British form. The understanding of the theoretical power and chemical implications of Newton's ideas which some members of this early group displayed was matched neither by their experimental ability nor by the existing state of chemical knowledge and technique. Though their theoretical disquisitions and practical investigations aroused considerable interest, they failed to carry through the chemical as opposed to the polemical program on which they embarked.

Later in the century the international prestige of Newton grew to such an extent that a physicalist program of reduction was more widely accepted among chemists. Early works of rather different style and content thus came to be read with admiration as guides to research within a more professionally

The Immediate Impact of Newton's Ideas

scientific and purely chemical context. Even so, Newtonian ideas on internal structure and quantified forces remained obstinately beyond the range of predictive chemical experience.

3.2. Gregory, Pitcairne, and the Beginnings of Discipleship

Intellectually the most eminent—though personally the most obnoxious—of the group round Newton was David Gregory.[3] A Scot, a mathematician, and an Edinburgh professor, Gregory came to London in 1691, met Newton and, largely through the latter's influence, was appointed to the Savilian chair of astronomy at Oxford. There he remained till his death in 1708, corresponding, visiting Newton in London, flattering, toadying, acting as focus for an active group of disciples centered on Christ Church, and keeping a gossipy and highly revealing private diary.

Newton's use of the preface to Gregory's 1702 *Elementa Astronomicae* as a pseudonymous vehicle for his own thoughts is characteristic, and well-known. An earlier example of the close intellectual relationship between the two men is seen in the set of notes that Gregory made about the as-yet-unpublished *Opticks* on a May 1694 visit to Cambridge. The following July he also correctly recorded—nineteen years before publication—that "in the new edition of Newton's philosophy . . . by far the greatest changes will be made to Book III. He will make a big change in hypothesis III page 402."[4] As the comment so graphically displays, many of Newton's ideas were known to his intimate friends long before their publication: a fact that immensely complicates any assessment of these friends' activities.

Gregory did much to encourage other Newtonians, including another widely influential early disciple, Archibald Pitcairne.[5]

---

[3] (1661–1708). Stewart (1901), *Academic Gregories*, provides a popular biography, while **Rigaud** (1838), *Essay on the Principia,* and Hiscock (1937), *Newton's Circle,* offer much incidental information.

[4] Turnbull (1959–61), *Newton Correspondence,* III, 338–339 and 384.

[5] (1652–1713). See Pitcairne (1715), *Life and Works.*

Atoms and Powers

Pitcairne, an Edinburgh physician, was first attracted to mathematics by Gregory's exposition. From iatromathematics to Newton was but a short step, and in March 1692 Pitcairne was in Cambridge, obtaining a manuscript version of the *De Natura Acidorum*.[6] Naturally enough Gregory was in due course to make his own copy of the manuscript from Pitcairne's original. Pitcairne's standing in the medical world was apparent in his 1692 election to the prestigious Leyden chair, and his unrepentantly iatromathematical approach from the lectures he delivered.[7] Though he remained in Leyden but a year, his visit was of great importance. It marked the start of a long and fruitful eighteenth-century connection between that university and Edinburgh.[8] Of more immediate consequence was the inclusion of such students as Richard Mead and Herman Boerhaave in his audience.

After his brief stay in Leyden, Pitcairne built up an extensive practice in Edinburgh. This practice did not prevent him being often in London, and in close contact not only with Gregory but later with such medical men and Newton-disciples as George Cheyne, Richard Mead, James Keill, and John Freind. Pitcairne's own involvement in Newtonian ideas is graphically displayed in a 1695 note to Gregory instructing him "to send me a copie of the notes on Mr. Newton's 2d. book. To give my most humble service to Mr. Newton, & to assure him that I'll put his Optics in Latin. To send me copies of all your lectures which you have written at Oxon. . . To endeavour to get Newton's papers about the mythologies; & Christian religion for me."[9] Pitcairne's surviving unpublished correspondence provides further fascinating insights into the Newtonian group, as in his 1703 injunction to "take notice that Dr. Gregorie and Dr. Cheyne are not indissoluble friends . . . [and] that Mr.

[6] Turnbull (1959–61), *Newton Correspondence*, III, 205–214.
[7] On the former, see Lindeboom (1963), "Pitcairne's Leyden interlude"; on the latter, Pitcairne (1715), *Life and Works*, and, for the opposition aroused, Pitcairne (1695), *Apollo Mathematicus*.
[8] On which, see Clow and Clow (1952), *Chemical Revolution*, chap. 25.
[9] Hiscock (1937), *Newton's Circle*, 3–4.

III. Pitcairne to Gregory in 1707. From Edinburgh University Library, Ms. Dc 1 62, f.158.

Craig is very far from being a friend to Dr. Gregorie. This for the politics. I am great with all." [10] (See also Fig. III.)

The developing sense of group awareness is also evident from Gregory's diary. For instance in Oxford in August 1704 Gregory recorded the gossipy comment that "Mr. Kyl [John Keill] told me that he saw the paper that Dr Cheyne gave to Dr Mead," while in London in January 1705 he was noting that "Dr Cheyne braggs that next summer he is to goe to Scotland, and together with Dr Pitcairn settle all the practice of physick, and publish unalterable principles thereof." And in an amusing note of 20 May 1706 he recorded how: "Dr. Cheyne uses to say among his chronys that all the additions (made by Sr Isaac Newton to his book of light and colours in the Latin version) were stolen from him by Sr Isaac in private conversation." [11] As this last note shows, the significance of Newton's 1706 Latin Queries was not lost on his contemporaries, even if it has sometimes escaped historians!

While Gregory and Pitcairne were of central importance in the establishment of a British Newtonian tradition, it is necessary to turn to lesser figures to see Newton's ideas on the nature of matter vulgarized, exploited, and propagandized for. That these lesser figures were almost all physicians is no accident, for until the late-eighteenth-century growth in chemical manufactures brought about drastic changes, chemistry was most often pursued as an adjunct to medicine. It is thus not surprising that the desire felt by Pitcairne and his followers to make medicine a Newtonian science should have had repercussions on chemistry. The other group most likely to express interest in the theory of matter and its chemical implications were those mathematicians who took an interest in natural philosophy. This latter group was represented in the early Newtonian circle not only by David Gregory but also by his able pupil John

---

[10] National Library of Scotland, Ms. 3440 consists of eight interesting letters from Pitcairne to the Rev. Colin Campbell. The quotation is from f. 20^v (a letter of 1 October 1703), by permission of the Trustees of the National Library of Scotland.

[11] Hiscock (1937), *Newton's Circle*, 19, 23, and 29.

Keill, who played such an active part in the encounter with Leibniz.

3.3. Cheyne, Mead, and the Medical Tradition

Born in Aberdeenshire, George Cheyne [12] was in due course attracted to the study of medicine by Pitcairne's Edinburgh teaching. His 1702 *New Theory of Fevers* plainly displays a Pitcairne-influenced iatromathematical approach, and the inspiration and encouragement afforded by Newton's work. Dismissing the still-influential "chymical physicians" and arguing boldly for a reductionist chemistry based on the new (Newtonian) physics, Cheyne could say: "All is nonsense, unless they first shew their systems and chymical effects to be necessary corollaries from the known laws of motion, i.e. unless all their philosophy, and chymistry too, be first mechanically explain'd." In his own words, the need was for a *Principia Medicinae Theoreticae Mathematicae,* the way forward having already been shown "by that stupendiously great man, Mr. Newton" who has provided "the only key, whereby the secrets of Nature are unlock'd." [13] With such sentiments, Cheyne was assured of election to the Royal Society and rapid entrance to the Newtonian circle when he moved from Edinburgh to London to practice.[14]

Not surprisingly, he was soon engaged in the swelling argument over the calculus, as well as in medical routine. Neither concern us here, but rather his 1705 *Philosophical Principles of Natural Religion*. To produce such a work was quite normal in an English context in which it was felt that "to discourse of [God] from the appearances of things, does certainly belong to natural philosophy." In passing, it may be noted that the maintenance of this connection between natural philosophy and Christian theology was one of the points that most sharply differentiated British from French thought, in both style and

---

[12] (1671–1743). See Mullet (1940), *Cheyne's Letters*.
[13] Cheyne (1702), 11, 24, and 27.
[14] See the note of introduction from Pitcairne that he carried with him: British Museum Sloane Mss. 4038 f. 246.

structure, throughout the eighteenth century. Even the growing professionalization of the physical sciences late in the century (itself a function of their relationship to manufacturing industry) seems at most to have weakened rather than ended this characteristically British connection.

The acknowledgments in Cheyne's work reveal the closeness of the circle round Newton at this time: "As for the materials... Some part of the matter was furnished me from Mr. Newton's store... For what concern'd the animal oeconomy... that friend to mankind, my constant good friend, Dr. Archibald Pitcairn his dissertations supply'd me the most plentifully... [I am also indebted to] the judicious corrections and advices, of ... Dr. Freind of Christ-Church Oxford." [15] The work itself is an early example of those popular expositions of "Newtonian" doctrine which we will consider in subsequent chapters. As such it reveals the sort of assumptions, unchallenged but undeveloped throughout the century, which take on fresh interest with the aid of hindsight. Cheyne asserted that "the particles of natural fluids must be similar, of equal diameters, of equal solidity, and consequently of equal specifick gravities, that they may be homogeneous, and of the same uniform nature... Water seems to consist of small, smooth, hard, porous, spherical particles of equal diameters, and equal specifick gravities." [16] To Cheyne though, as to all contemporary Newtonians, such ideas were of little interest—for true chemical understanding, as opposed to mere classification, lay in the pursuit of measurable forces and the exploration of "more secret and noble" internal structures.

The sort of Newtonianism advocated by Cheyne may also be seen in the works of Richard Mead,[17] another highly successful physician of the period. Though he received his medical education on the Continent, he also encountered Pitcairne: "After three years residence in Utrecht, Mr. Mead went to Leyden, and ... attended ... the lectures on the theory and practice of

---

[15] Cheyne (1705), sigs. A6 and A7.
[16] *Ibid.*, 61.
[17] (1653–1754). See Mead (1755), *Authentic Memoirs*.

The Immediate Impact of Newton's Ideas

physick by the famous Pitcairn."[18] Mead too became a disciple, as is clear implicitly from the style and explicitly from the preface of his 1704 work, *De Imperio Solis ac Lunae in Corpora Humana*.

Mead, who was eventually to become Newton's own physician, was admitted F.R.S. on the very day that Newton assumed the presidency. Mead's election was no doubt assisted by the 1702 *Mechanical Account of Poisons*, which made quite plain his mathematical bias ("it may be hoped in a short time . . . that mathematical learning will be the distinguishing mark of a physician from a quack"), his approval of Cheyne ("[who] has enumerated several particulars in which the theoretic part of medicine still wants improvement"), and his great admiration for Newton's work on the attraction of particles to one another ("this Mr. Newton has demonstrated to be the great principle of action in the universe"). Mead, not unnaturally, had little time to go into chemical discussion, and could only pause to assert that

> he who rightly studies [Newton's] philosophy will understand, that the same [principle of action] obtains in the most minute and finest corpuscles . . . And hence it follows, that whatsoever power is sufficient to make a change in this attraction, or cohesion of the parts, makes an alteration of the nature of the fluid; that is, as the chymists express it, puts it into a fermentation.[19]

To see the chemical implications of Newton's work fully developed (on the theoretical level), it is necessary to turn to other members of this group, especially the brothers John and James Keill and "the famous Dr. Freind" with whom Mead "lived in the strictest intimacy."[20]

[18] *Ibid.*, 5.
[19] Mead (1702), sigs. a6^{r,v} and 15.
[20] Mead (1755), *Authentic Memoirs*, 22. Something of Mead's later importance as intelligencer and propagator of Newtonianism is apparent from the same account (pp. 55–59): "His reputation . . . as a scholar was so universally established, that he corresponded with all the principle Literati of Europe . . . No foreigner of any learning, taste, or even curiosity, ever came to London without being introduced to Dr. Mead." See also note 44 to Chapter 4, below.

3.4. The Keills, Freind, and Leibnizian Warfare

The group more or less closely clustered round Newton in early-eighteenth-century London fulfilled a variety of purposes. While enlarging the bounds of natural philosophy was clearly one of these, promoting a particular style and metaphysic, and defending the national honor, were equally clearly others. And in such a politically troubled time, with the very succession to the throne insecure, these latter aspects were central to discussion and debate. That there were profound metaphysical differences between Newton and Leibniz, the two greatest mathematicians and natural philosophers of the age, was unfortunate enough. That they independently arrived at two different formulations of the calculus was still more so. But that Leibniz alone had the ear of the Electress Sophia, the destined English monarch, was—in light of Newton's neurotic, suspicious, and jealous character—little short of a disaster.

The dispute over the discovery of the calculus was a complex, tangled affair, which reflects little credit on any of the parties involved. Beginning as a purely technical quibble, it rapidly escalated into a full-blooded debate over the whole scope, style, and structure of natural philosophy, a debate that culminated in the *Leibniz-Clarke Correspondence*. In a sense, Newton won. At least the alarmed notes in the Royal Society's *Journal Book* for 11 November 1714 that Leibniz intended to publish his own "Commercium Epistolicum in a defence of himself at his return from Vienna to Hanover," and that—even worse—"Mr Leibnitz designed in a short time to be in England," [21] proved unfounded. The prediction, if not the language, of John Keill's statement that "Mr. Leibnitz . . . will not have the impudence to show his face in England" [22] turned out to be more accurate.

---

[21] *Journal Book of the Royal Society (Copy)*, XI, 25, quoted by permission of the President and Council of the Royal Society of London. The *Journal Book* will be cited subsequently as *JB*.

[22] Quoted from a 6 August 1714 letter to Newton: Cambridge University Library, Ms. Add. 3985 f. 13 ʳ.

The Immediate Impact of Newton's Ideas

Even before Leibniz' untimely death, his royal patron seems to have sensed the political winds enough to acknowledge Newton's claim to rule British natural philosophy without challenge.[23]

The dispute itself, especially as it relates to the calculus and to broad metaphysical issues, cannot be treated here. Yet precisely because it was in the context of this dispute that the Newtonian "nut-shell" theory of matter was fully elaborated and the program of force-measurement spelled out, the controversy did profoundly affect the course that was followed by chemical research and debate. Not only was the Newtonian position defended, and campaigned for, with an enormous zeal. The battle with Leibniz was also one reason why French philosophers were if anything more hostile to Newton's work in 1725 than a generation earlier, and why German chemists were to be so conspicuously absent from the later Newtonian research program.

3.4.1. The "Nut-Shell"[24] Theory

Though Newton had long brooded over the nature of the internal structure of matter, he did not choose to make his thoughts public before 1706. The first edition of the *Opticks* was content to state that "bodies are much more rare and porous than is commonly believed ... And he that shall find out an hypothesis, by which water may be so rare, and yet not be capable of compression by force, may doubtless by the same hypothesis make gold, and water, and all other bodies, as much rarer as he pleases; so that light may find a ready passage through transparent substances."

A similar modesty was apparent in such Newtonian-inspired texts of the period as John Keill's *Introductio ad Veram Physicam*. As a mathematician, astronomer, and pupil of David Greg-

---

[23] The important and so-far neglected political motivation of the Newton-Leibniz quarrel is explored, and fuller documentation provided, in Thackray (in press) "The business of experimental philosophy."

[24] The phrase comes from Priestley (1777), *Matter and Spirit*, 17; the present discussion is based on Thackray (1968b), "Matter in a nut-shell," which provides a full account of this theory and its eighteenth-century influence.

ory, Keill [25] had moved to Oxford with his master, and thus was early part of the developing Newtonian group. In 1700 he delivered a course of university lectures based on the *Principia* and designed to serve as introduction to it. In contrast to what would appear only six years later, the 1702 published version of these lectures was a model of restraint and caution. Not only did it scarcely mention forces but on the subject of rarity and internal structure it merely stated that "he who has understood what was said in these lectures, can easily show the solution to the following problem." The problem was to prove that a grain of solid matter might be so dispersed as to fill any finite space, however large, without leaving pores greater than any given size. No proof was offered. Instead Keill concluded by saying that "in the same way as this problem is solved it can be demonstrated how it is possible for one grain to render the whole of that space dark and prevent the transmission of even the smallest ray of light." [26]

John Harris,[27] lexicographer, coffeehouse-frequenter, Newtonian, and popular lecturer, was more cautious still. The 1704 volume of his *Lexicon Technicum* reproduced the *Principia* argument for the necessity of a disseminate vacuum. Yet it did not so much as hint at a "nut-shell" theory of matter (a theory, that is, which downgrades the role of brute matter in Nature to the point of supposing that—in Joseph Priestley's vivid phrase —"all the solid matter in the solar system might be contained within a nut-shell.") Even the reviews of Keill's book and of the *Opticks* in the Leibniz-favoring *Acta Eruditorum* expressed no unease.[28] Instead the intensifying clash over calculus-priority was the main concern. The 1704 *Opticks* had not limited itself to the phenomena of light. Nor had it stopped at points associated with Newton's view of the natural world as an almost

---

[25] (1671–1721). See Hutton (1795), *Philosophical Dictionary*.

[26] Keill (1702), *Introductio*, 55.

[27] (1667?–1719). See Taylor (1954), *Mathematical Practitioners*. See also section 8.1. below for a discussion of Harris as popular lecturer.

[28] See Harris (1704), article "vacuum"; *Acta Eruditorum* (1703), 504–510 and (1706), 59–64.

The Immediate Impact of Newton's Ideas

vacuous entity, operated on by immaterial forces sustained by divine decree. It had also initiated the serious warfare of the calculus-priority dispute. How acutely aware of this aspect of the work were Newton's own immediate circle, and how closely intertwined were philosophy and politics, may once again be glimpsed from Pitcairne's correspondence. In October 1703 he noted that "2 sheets [of the *Opticks*] are cast off already. He [Newton] adds to this book all his about quadratures. He has done it in ire, being barbarously . . . and hanoverianlie abused for his *Principia* by a German latelie in a vile consubstantial book of nonsense and ill-nature." [29]

Leibniz was at pains to rebut the charge of plagiarism implied in the preface and mathematical appendices to the *Opticks*. It would seem that this rebuttal strengthened Newton's determination to mount a full-scale attack on both Leibniz' mathematical claims and his philosophy. With Leibniz as court philosopher of the heir to the English throne, Newton, the newly established ruler of British science, had more than narrowly scientific incentives to do battle with a natural philosophy so strongly opposed to his own in assumptions and conclusions. To be "hanoverianlie abused" was a matter of concern at such a moment. The 1706 *Optice* was therefore fortified by a number of additional Queries, setting out more fully the role of attractive forces in the economy of Nature. At the last minute Newton also added two fresh pages to the main body of the work.

These pages developed a "hypothesis" to explain his 1704 statement that "bodies are much more rare and porous than is commonly believed." The explanation given related to optics, but was of much wider implication. Two things make this clear. One is that this same hypothesis occurred in a suppressed 1690's draft for the *Principia*. The other is the way David Greg-

---

[29] Quoted from a 1 October 1703 letter to the Rev. Colin Campbell: National Library of Scotland, Ms. 3440 f. 20. On the calculus dispute, see Thackray (in press), "The business of experimental philosophy"; Manuel (1968), *Portrait of Newton*, chap. 15. The "vile consubstantial book" was presumably Johannes Gröning's *Historia Cycloeidis*, published in Hamburg in 1701. See also Hiscock (1937), *Newton's Circle*, 26–27.

ory recorded his reaction to the hypothesis. In a December 1705 memorandum, Gregory noted how "one of the great scruples men commonly have about matter's being proportioned to weight, is that water is more than 12 times lighter than gold, & gold scarce half full, as appears by many experiments; thus water is not one part of twenty full matter, & yet by the strongest machine incomprehensible. Sir Isaac Newton proposes this as one theory of making bodys in any degree porous & yet solid." Then followed a delightfully clear summary of the "nut-shell" theory, as inserted in the *Optice*. Yet so hesitant was Newton about publishing this "hypothesis," that it appeared not in the main text, but only tucked away in small print in the corrigenda at the front of the book.[30]

That the hypothesis emerged in print at this particular moment seems to have been principally a function of the growing dispute with Leibniz. Not only did the hypothesis serve to emphasize the breach between Leibnizian (and Cartesian) plenum theories and Newton's beliefs, and to magnify God by minimizing the role of matter in Nature, it also—by stressing infinite divisibility rather than indivisibles—indicated how Newton's method of fluxions was all of a piece with his natural philosophy.

Despite these various advantages, Newton characteristically preferred to see the "nut-shell" idea develop through the writings of his disciples, rather than say more about it himself. In 1708 a threefold attack was launched by these disciples. Leibniz' claim to mathematical priority was again disputed, a vigorous program of explanation by attractive forces begun, and the contrast between any plenist-materialist universe and Newtonian belief in vacuity and powers actively developed. Only this last point directly concerns us here, but the rest of the attack must

---

[30] Newton (1706), sigs. b 1–2. The Gregory quotation is from Hiscock (1937), *Newton's Circle*, 29–31, where Gregory also records a mathematical formulation of the "nut-shell" theory. For the *Principia* draft, see Hall and Hall (1962), *Unpublished Newton*, 312–317. See also McGuire (1966), "Body and Void," note 16. For Leibniz' reaction to Newton's public statements of 1704, see *Acta Eruditorum* (1705), 30–36.

The Immediate Impact of Newton's Ideas

necessarily be touched on, not least because the calculus-priority assertions and the "nut-shell" theory both hinged on a denial of those indivisibles on which the Leibnizian calculus depended. Thus—paradoxically—Newtonians were increasingly committed to a belief in the infinite divisibility of matter at the very time when Newton was also moving toward a public declaration of his own life-long belief in atoms!

The 1708 *Philosophical Transactions* contained two papers by John Keill.[31] One touched on calculus-priority. The other displayed the universal rule in Nature of attractive forces, by means of thirty interlocking theorems. The first of these, again offered without proof, was the one that appeared in the 1702 *Introductio*. This time it was backed up by two further (unproven) theorems. One stated how, for instance, though the quantity of matter in a cubic inch of gold "may be 20,000 times as great" as that in a cubic inch of air, "yet the vacuities in the gold, may be to those in the air, as 999,999 to 1,000,000, i.e. very nearly equal." The other explained how "those particles which constitute water . . . are not absolutely solid; but are compounded of other particles, which do contain within them many pores or vacuities."[32] Though only briefly amplified, and not dignified as such, this was the "nut-shell" theory of matter in all its essentials. And that the import of this statement was not lost on contemporaries is plain from the *Commonplace Book* of the future Bishop Berkeley.[33]

The same themes were echoed in James Keill's *Animal Secretion*. Once more the "nut-shell" theory was implied but not stated. The main space was reserved for the role in Nature of attractive forces, and Newton was again subject to a eulogy,

---

[31] Keill (1708 b and c). Cf. *JB*, X, 195: "November 3d 1708. The President in the Chair. Severall theorems of Mr Keill were shown which were judged fitt to be printed in the Transactions." The number and variety of such references in the *Journal Book* reveal how Newton was actively aware of, and gave his blessing and oversight to, those activities of the early Newtonians with which we are here concerned.

[32] Keill (1708b), 98. Translation quoted from Harris (1710), *Lexicon Technicum*, article "particle."

[33] Johnston (1930), 43. See section 8.3. below.

especially for the Queries newly appended to the *Optice*. In 1709 the bombardment was further reinforced with John Freind's *Praelectiones Chymicae* and Francis Hauksbee's *Physico-Mechanical Experiments*.[34] Finally John Harris' 1710 second volume to the *Lexicon Technicum* provided a convenient summary of the whole Newtonian system, particularly "of that most amazing property, the attraction of the particles of matter" and of what "those ingenious and industrious brothers, the Keills, have publish'd about this affair of attraction." For good measure, a first printing of Newton's *De Natura Acidorum* (which also dealt with short-range forces and internal structure) was included. And just in case the significance of all this should escape anyone, Harris was careful "further to inform the reader, that there is now printing a Latin mathematical treatise or two, of Sir Isaac Newton's, which were written many years ago, and which by their date, will sufficiently determine, whether the new methods of fluxions were known first to him, or Mr. G. Leibnitz." [35]

The Leibnizians were not slow to realize the seriousness of all this. In September 1709 the *Acta Eruditorum* carried a hostile review of James Keill's book, while in 1710 Leibniz' own *Theodicy* appeared: this latter contained a sustained attack on the occult nature of attractive forces. In January 1710 the *Acta* also belatedly printed an article by Christian Wolff criticizing the arguments for a vacuum made use of in Keill's *Introductio*. A

---

[34] Keill (1708a). See pp. 12–13 for the "nut-shell" implication, and pp. 8–9 for Newton and "the questions annexed to the Latin edition of his *Opticks*." See also p. xxv on how short-range forces were "the doctrine of Hippocrates." See also Freind (1709), Dedication and pp. 2–3, and Hauksbee (1709), 156–158. Cf. *JB*, X, 207: "March 9th 1709. . . . An account of Doctor Freind's *Praelectiones Chymicae* was read, as likewise an account of Doctor Keill's *Treatise of Animal Secretion*."

[35] Harris (1710), introd. At the time of producing this second volume, Harris was secretary of the Royal Society, and thus in immediate contact with Newton, its president. That the *De Natura Acidorum* appeared on Newton's own initiative is clear from *JB*, IX, 234: "February 15th 1710. . . . A paper of Sir Isaac Newton's concerning acids was read. It was ordered to be printed in the *Transactions*." Instead it appeared in Harris's 1710 volume. The contrasts between this volume and its 1704 companion would make a rewarding study in the development of Newtonianism.

prompt reply by Keill evoked only further criticisms from Wolff. The September *Acta* carried a further and much more hostile article, this time based on Freind's *Praelectiones*. And in March 1711 Leibniz made formal complaint to Hans Sloane about the charges of plagiarism.[36] Small wonder then that in April 1711 John Keill, writing to Newton, should say: "I wish you'd take the pains to read that part of [the *Acta*] supplements wherein they give an account of Dr. Freind's book and from thence you may gather how unfairly they deal with you." Small wonder either that the *Philosophical Transactions* was soon printing a long reply by Freind,[37] or that Leibniz was pressing home the attack in his public correspondence with Hartsoeker.

Most of these earlier exchanges were ostensibly concerned either with the calculus, or with the admissibility and utility of the term "attraction." We shall return to this latter aspect below, but here it is necessary to concentrate on the fortunes of the "nut-shell" theory of matter. Faintly suggested in Keill's *Introductio* and the 1704 *Opticks,* developed only in the corrigenda of the *Optice,* and implied rather than stated in the 1708 theorems and Harris' 1710 volume, the theory was still sufficiently peripheral to pass unscathed in the Leibnizian reviews. The alterations that Newton introduced into the long-awaited second edition of the *Principia,* the appearance of the *Commercium Epistolicum,* and the further activities of John Keill, were to change this state of affairs.

When the second edition of the *Principia* finally apeared, it contained major alterations to the section on matter and the vacuum at the start of Book III. These changes made much clearer Newton's commitment to a vacuum, and to the infinite divisibility of matter. On this latter point the "Rules of reason-

---

[36] *Acta Eruditorum* (1709), 24–31 and 397–404; (1710), 11–15, 78–80, 412–416. See also Cohen and Koyré (1962), "Leibniz-Clarke Correspondence." The draft of Keill's 1710 reply to Wolff is still extant, in packet no. 7 of the (uncatalogued) Lucasian Mss. in Cambridge University Library.

[37] Cambridge University Library, Ms. Add. 3985 f. 1^r, and Freind (1712 a and b). See Thackray (1968b), "Matter in a nut-shell," note 17, for further information on this particular exchange.

ing in philosophy" were careful to point out that "had we the proof of but one experiment that any undivided particle, in breaking a hard and solid body, suffered a division, we might by virtue of this rule conclude that the undivided as well as the divided particles may be divided and actually separated to infinity."[38]

To appreciate the relationship of this statement to the attacks on Leibniz, we must turn first to Newton's own (anonymous) *Philosophical Transactions* review of the *Commercium Epistolicum*. This review stated quite unequivocally that "indivisibles, upon which [Leibniz'] differential method is founded, have no being either in geometry or Nature."[39] If we then turn to a long article in the April–June 1714 issue of the *Philosophical Transactions,* the developing Newtonian position on the infinite divisibility, rarity, and tenuity of matter will be plain. The second article appeared under John Keill's name and took the form of two mathematical theorems. Ostensibly arising from his 1710 exchange with Wolff, they extended Keill's 1702 and 1708 remarks into the first formal elaboration of that "nutshell" theory of matter which was to ensure the link between fluxions and physical reality (or—more accurately—to deny the link between the same reality and the Leibnizian calculus).[40]

The appearance of Keill's theorems can only have served to

[38] Newton (1713), 358. Translation quoted from Newton (1729), II, 204.

[39] Newton (1715), 205. This statement was to enjoy much subsequent repetition by British Newtonians. See e.g., Wilson (1761), *Robins' Mathematical Tracts*, II, 76–80: "Herein the method of prime and ultimate ratios essentially differs from that of indivisibles; for in the method of indivisibles momenta are considered absolutely as parts, whereof their respective quantities are actually composed . . . [But this being unintelligible] Sir Isaac Newton therefore instituted a manner of conception on quite different principles. He observing (to use his own words) that indivisibles have no being either in geometry or in Nature . . ." For a further example of how interlocked were the various aspects of the Newton-Leibniz quarrel, see Newton (1715), 224. Here Newton claims that by identifying attraction with Henry More's "hylarchic principle," the Leibnizians "are persuading the Germans that Mr Newton wants judgement, and was not able to invent the infinitesimal method." See also Cohen and Koyré (1962), "Leibniz-Clarke Correspondence," 63–68.

[40] Keill (1714). See also the somewhat confused discussion of Newton, Keill, and infinite divisibility, in Strong (1951), "Newton's mathematical way," and (1957), "Newtonian explications."

The Immediate Impact of Newton's Ideas

increase the mounting anger of Leibniz. In November 1715 that anger erupted in a letter to Princess Caroline, his friend and former pupil. What was to become the *Leibniz-Clarke Correspondence* opened with his complaints about how "natural religion itself seems to decay [in England] very much," and how "many will have human souls to be material," and in particular how "Sir Isaac Newton, and his followers, have also a very odd opinion concerning the work of God."

The reply from Samuel Clarke, theologian and apologist for Newton, was quick to agree "that some make the souls of men and others even God himself to be a corporeal being; is also very true, but those who do so, are the great enemies of the mathematical principles of philosophy, which principles, and which alone, prove matter, or body, to be the most inconsiderable part of the Universe." The Newtonian doctrine of the almost "matterless" nature of the universe was thus part of the discussion from its early stages. The ensuing exchange of letters ranged widely, involving the Leibnizian principles of continuity, sufficient reason, and perfectability, as well as God's mode of operation in Nature and the actual properties of matter.

In his second letter Leibniz showed himself well aware of the "nut-shell" aspect of Newton's view of matter, and stated the theological basis of his objection to such beliefs:

According to Sir Isaac Newton's philosophy . . . matter is the most inconsiderable part of the universe . . . according to his notions, matter fills up only a very small part of space. But Democritus and Epicurus maintained the same thing; they differ'd from Sir Isaac Newton only as to the quantity of matter; and perhaps they believed there was more matter in the world than Sir Isaac Newton will allow: wherein I think their opinion ought to be preferred; for, the more matter there is, the more God has occasion to exercise his wisdom and power. Which is one reason, among others, why I maintain there is no vacuum at all.

Atoms and Powers

Characteristically, Clarke's reply asserted the impeccable "Phoenician" ancestry of belief in matter and a vacuum, while denying the validity of Liebniz' theological objections. Nothing daunted, Leibniz repeated these objections in his much lengthier third lettter. In reply, Clarke stuck to his own position that the "determinate quantity of matter that is now in the world is the most convenient for the present state of Nature." [41] The further letters on either side continued the dispute, but contained no significant additions to the argument over tenuity.

From what has been said, it should be apparent that the "nutshell" view of matter became just as much part of the Newtonian position, as belief in a plenum was of the Leibnizian. That Newton took the whole exchange seriously enough to contemplate further revising the opening sections of Book III of the *Principia* in its light is scarcely surprising.[42] These revisions were never executed, perhaps because the eventual production of a *Principia* third edition was a work of Newton's extreme old age. The *Opticks* fared better. The second English edition was published in 1717, hot on the heels of the Leibniz-Clarke controversy, and shortly after Leibniz' death.

In this new edition, the section of Book II, part III, that dealt with bodies being "much more rare and porous than is commonly believed" contained a slightly modified version of the two pages of further discussion that first appeared in the 1706 corrigenda (see Figs. IV and V). Now at last Newton permitted himself to state publicly, as an unavoidable part of his text, the "nut-shell" theory that David Gregory had so carefully noted down from his lips, twelve years before. To quote at length from the 1717 *Opticks* version:

[41] The correspondence is best consulted in the edition by Robinet (1957), which gives information on the various personnel. Thackray (in press), "The business of experimental philosophy," provides references to the voluminous historical literature on the more narrowly political relationships between London and Hanover. The quotations are cited from the English edition of the *Correspondence* by Alexander (1956), 11, 12, 16.

[42] The proposed revisions and their significance are extensively discussed in McGuire (1966), "Body and void."

The Immediate Impact of Newton's Ideas

[70]

other Bodies as much rarer as he pleases, so that Light may find a ready passage through transparent substances.

PROP. IX

Bodies reflect and refract Light by one and the same power variously exercised in various circumstances.

This appears by several Considerations. First, Because when Light goes out of Glass into Air, as obliquely as it can possibly do, if its incidence be made still more oblique, it becomes totally reflected. For the power of the Glass after it has refracted the Light as obliquely as is possible if the incidence be still made more oblique, becomes too strong to let any of its rays go through, and by consequence causes total reflexions. Secondly, Because Light is alternately reflected and transmitted by thin Plates of Glass for many successions accordingly, as the thickness of the Plate increases in an arithmetical Progression. For here the thickness of the Glass determines whether that power by which the Glass acts upon Light shall cause it to be reflected, or suffer it to be transmitted. And, Thirdly, because those surfaces of transparent Bodies which have the greatest refracting power, reflect the greatest quantity of Light, as was shewed in the first Proposition.

PROP. X.

If Light be swifter in Bodies than in Vacuo in the proportion of the Sines which measure the refraction of the Bodies, the forces of the Bodies to reflect and refract Light,

IV. Newton revises the *Opticks*. From Newton's copy of the 1704 edition, in Cambridge University Library.

How bodies can have a sufficient quantity of pores for producing these effects [transmission of light etc.] is very difficult to conceive, but perhaps not altogether impossible. For . . . if we conceive these particles of bodies to be so disposed

63

Atoms and Powers

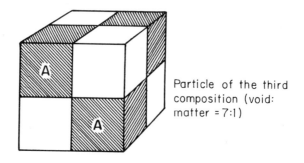
Particle of the third composition (void: matter = 7:1)

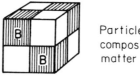
Particle of the second composition (=A. Void: matter = 3:1)

Particle of the first composition (=B. Void: matter = 1:1)

 Ultimate particle or atom (=C. No void space)

V. The internal structure of matter.

amongst themselves, that the intervals or empty spaces between them may be equal in magnitude to them all; and that these particles may be composed of other particles much smaller, which have as much empty space between them as equals all the magnitudes of these smaller particles: And that in like manner these smaller particles are again composed of others much smaller, all of which together are equal to all the pores or empty spaces between them; and so on perpetually till you come to solid particles, such as have no pores or empty spaces within them: And if in any gross body there be, for instance, three such degrees of particles, the least of which are solid; this body will have seven times more pores than solid parts. But if there be four such degrees of particles, the least of

The Immediate Impact of Newton's Ideas

which are solid, the body will have fifteen times more pores than solid parts. If there be five degrees, the body will have one and thirty times more pores than solid parts. If six degrees, the body will have sixty and three times more pores than solid parts. And so on perpetually.

Having said all of which, the passage enigmatically concluded: "and there are other ways of conceiving how bodies may be exceeding porous. But what is really their inward frame is not yet known to us." [43]

There is no reason to doubt Newton's sincerity of intention in this last remark. On the one hand his care in distinguishing between possible explanations and demonstrated truths was a major aspect of his unrivaled scientific ability. On the other, polemical need of infinite divisibility sat ill with his deep emotional commitment to "the Ancients" and their atomism. Yet his reiteration of this "conception" of the "inward frame" of bodies in other published passages, the tenor of a number of his manuscripts, the 1708 and 1714 statements of John Keill, the course taken by the *Leibniz-Clarke Correspondence,* all leave us with little room for doubt about the importance of the "nut-shell" idea in Newton's own thought. Certainly the commentators had no such doubt. With Leibniz dead, and Newton himself at last plainly committed to an extended statement of the "nut-shell" theory, what need had they of further caution?

The changes that John Keill made for the third edition of his textbook were indicative of the new Newtonian orthodoxy. In place of the oblique hints of the 1702 and 1705 printings, this 1715 edition of the *Introductio* included the 1714 mathematical proof of the theorem, first baldly stated in 1708, about how a grain of solid matter might be made to fill any finite space. The "nut-shell" implication was driven home in the corollary that "there may be given a body, whose matter if it be reduced into a space absolutely full, that space may be any given part of the former magnitude." The second 1708 theorem, about how two

[43] Newton (1718), 242–244.

bodies of equal bulk but widely differing quantities of matter might contain almost equal amounts of empty space, was also reproduced in its full 1714 form. As in 1714, Keill was not afraid to say "perhaps the quantity of matter in glass has not a greater proportion to its magnitude, than a grain of sand to the whole bulk of the globe of the earth." Neither was he slow to point out the explanations the "nut-shell" idea could offer. These included reasons as to "why [shades of Newton!] the magnetick effluvia pervade with the same facility both dense gold and the more rare air," and "why the rays of light proceeding from many different objects, and transmitted through a small hole, do not mutually hinder one another." [44]

A host of later commentators took their clue from Keill. Already in 1720, W. J. 'sGravesande was careful to include Keill's first theorem in his highly successful *Physices Elementa*. It fitted well in such a Newtonian text. Naturally both Keill himself and Desaguliers kept the theorem in their rival translations of the work.[45] And how actively the minds of Newtonians were running on the "nut-shell" question is apparent once more from the Royal Society's *Journal Book*. A June 1720 entry shows "the President in the Chair," and the faithful Desaguliers producing an experiment designed "to disprove the Cartesian hypothesis of a plenum of matter."

Again it is instructive to look at Henry Pemberton's best-selling *View of Sir Isaac Newton's Philosophy*. This "authorized" popularization of the *Principia* and *Opticks* carefully repeated the 1717 argument about the internal structure to matter. And, following Keill, Pemberton *explicitly* stated that "this whole globe of earth, nay all the known bodies in the universe together, as far as we know, may be compounded of no greater portion of solid matter, than might be reduced into a globe of one inch only in diameter, or even less." Thus what had origi-

---

[44] Keill (1715), 56–59. Translation quoted from Keill (1720), *Natural Philosophy*, 65–67.

[45] See 'sGravesande (1720–21), I, 12. Cf. the English translations: 'sGravesande (1720a), I, 10, (1720b), I, 12. See also note 32 of Chapter 4, below.

The Immediate Impact of Newton's Ideas

nally been expressed only as a possibility—though admittedly a possibility close to Newton's heart—was by the 1730's well on the way to becoming an unquestioned assumption.[46] As we shall see in subsequent chapters, belief in the great paucity of matter in the universe and in a complex internal structure to all known chemicals, underlay much later-eighteenth-century thought. Thus a belief polemically developed in one science, profoundly affected the course of another.

3.4.2. The Theoretical Discussion of Forces

We have already indicated the controversial intent that lay behind the 1706 Queries, and the associated salvo of Newtonian works on short-range forces. Nonetheless it will be worthwhile to examine the works more closely, for their reasoning was ingenious and persuasive. While this reasoning seems to have had little effect on Leibnizians save to enrage them, it did powerfully influence later chemists seeking to understand and quantify chemical forces, and grateful for models on which to base their work.

In this connection Keill's 1708 theorems were widely influential. It is characteristic that although, by his own admission, he had been working on these theorems for five years previously, they were published only after the *Optice* had appeared, and then not so much in their own right but as part of a sustained campaign on behalf of attractive and repulsive forces. Keill's theorems considered particles that attract each other by another power, besides that of gravity, with this other power diminishing at a greater rate than the inverse square of distance. He showed that if this force had a finite ratio to gravity at the point of contact, it would vanish at all assignable distances. The later theorems then went on to demonstrate how such a short-range force might be used to explain cohesion, fluidity, and the cause of elasticity and softness.

---

[46] Pemberton (1728), 356. As late as 1788, such an influential British writer as William Nicholson was to reproduce Keill's theorems in their entirety, and argue at length for the paucity of matter, on the basis of the *Opticks*: see Nicholson (1788), *Natural Philosophy*, 24–26.

Keill further extended this idea of an additional interparticle attractive force besides the gravitational force to explain the solution of salts, fermentation, effervescence, crystallization, freezing, and precipitation, as well as the phenomena of electricity. Indeed his explanations show a remarkable similarity to those Boscovich was to put forward half a century later, in a far more developed and sophisticated form. And interestingly enough, they also show a remarkable similarity to Newton's own unpublished speculations.[47]

The attempt to apply Newtonian explanations to chemical phenomena also engaged Keill's younger brother James.[48] With John's help and advice, he sought to apply the principle of attraction in explaining animal physiology. Like Mead and Cheyne, James Keill was not only of Scottish descent but also a physician. Author of a highly successful *Anatomy of the Human Body,* he also translated the later editions of Lemery's *Cours de Chymie,* and so was well aware of the older "corpuscular" approach to chemistry. Once again, his shift in allegiance seems to have been gradual but thorough, so that in spite of his earlier sympathies, James was willing and eager to join in the post-1706 Newtonian barrage.

His *Animal Secretion* clearly reflects his involvement with both the personnel and ideas of the Newtonian circle. As regards the one, we read how "Dr. Pitcairne has explained the mechanical structure of the lungs . . . Dr. Freind has wrote in a mechanical way upon the menses; Dr. Cheyne upon fevers; and Dr. Mead of poisons." Of the other, we learn not only that (following "the learned Dr. Gregory") "the gravitation of the heavenly bodies towards one another was known to the ancient philosophers" but also that the rather different "power by which small particles of matter attract one another was the doctrine of Hippocrates." And should anyone feel this proof by citation of ancient authority dubious or insufficient, James, betray-

---

[47] Keill (1708b). Cf. Newton (1687), *Principia,* Bk. I, props. 85 and 86; see also Newton's 1690's speculations about how "the particles of bodies have certain spheres of activity," reproduced in Cohen (1966), "Newton's philosophy," 179.

[48] (1673–1719). See *DNB.*

The Immediate Impact of Newton's Ideas

ing all, resorts to the familiar assertion that "such an attractive power in Nature as this we have mentioned . . . can be denied by none, that duly consider the experiments and reasons given for it, by Sir Isaac Newton, in the Questions annexed to the Latin edition of his *Opticks.*" In a further interesting aside, James also admitted that the idea of a short-range attractive force was "first communicated to me by my brother at Oxford, above seven years ago; who had no sooner discovered it, but he deduced from it the cohesion of the parts of matter, the cause of . . . fermentations, dissolutions, coagulations, and many other of the operations in chymistry." [49] Once again, the dominance of Newton is displayed. For if this statement is correct, John Keill, who made no public use of them until 1708, was yet aware of and developing Newton's ideas on short-range forces from about 1700!

Animal Secretion is a work of great interest, not only for the history of physiology and of internation scientific strife, but also because of the theoretical insights in contains. Using the unarticulated assumption that the short-range force, like the gravitational force, is always in the same ratio to the inertial quantity of matter, Keill can say: "A large particle attracts not more strongly than a small one of the same solidity, but a diversity of figures causes different degrees of attraction in particles, that are otherwise the same . . . so that the attractive force varies, according as the particles are cones, cylinders, cubes, or spheres, and *caeteris paribus* a spherical particle, has the strongest attractive power." [50] This passage clearly states an idea implicit in Newton's work, and one to be of increasing importance through the century. The belief that the efficacy of the short-range attractive force was governed by particle type came to condition much research aimed at quantifying chemical mechanisms, especially when Boscovich and Buffon further developed the idea in their different ways.

Keill's book is chemically interesting, not only for these dim

[49] Keill (1708a), iv, xxv, 7–9.
[50] *Ibid.*, 18–19.

foreshadowings of what lay ahead, but also for a passage which illustrates how close in substance but far in spirit from nineteenth-century formulations such men were. Concerned to indicate how a force-based system and a few simple particles ("cones, cylinders, cubes, or spheres") could explain the formation of gall, seed, bile, etc. from the blood, he argued that

> a few different sorts of particles variously combined, will produce great variety of fluids, some may have only one sort, some two, some three, or more ... If we suppose only five different sorts of particles in the blood, and call them a, b, c, d, e, their several combinations, without varying the proportions, in which they are mixt will be these following.
>
> | ab: | ac: | ad: | ae: |
> |---|---|---|---|
> | bc: | bd: | be: | cd: |
> | ce: | de: | abc: | adc: |
> | bdc: | bde: | bec: | dec: |
> | $abcd$: | $abce$: | $acde$: | $abde$: |
> | $bcde$: | $abcde$. | | |
>
> But whether there are more or fewer in the blood, I shall not determine.[51]

Such speculations reveal the power of an algebraic approach to chemistry. They also show the fatal weaknesses of so much early Newtonian work: its lack of firm basis in experiment, and its ability to verbalize each and every observation.

John Keill's general theorems on attractive forces, and James's application of them to the particular case of the human blood, were enough to excite the hostility of Leibnizians: how much more so John Freind's thoroughgoing attempt to reformulate a complete science on Newtonian principles! As a Student of Christ Church from 1694, Freind [52] soon came under the influence of Newtonian enthusiasts. In 1703—the year he graduated M.D.—he published his *Emmenologia,* in which menstruation was explained "on mechanical principles." Com-

---

[51] *Ibid.,* 61–62.
[52] (1675–1758). See Nichols (1817–58), *Literary History,* V, 93–103.

The Immediate Impact of Newton's Ideas

ing well before the *Optice,* this work naturally made no reference to attractive forces. Instead it contained a series of "mechanical" propositions reminiscent of the *Principia*. The contrast with the 1709 *Praelectiones Chymicae* is marked. This latter work (supposedly the text of a 1704 university course of chemical lectures) was dedicated to Newton, with direct and pointed reference to the multiplicity of experiments on attractive forces contained in the *Optice*.[53]

The *Praelectiones* are of great interest both for their thoroughgoing reductionism, and for the stormy controversy they provoked with "the Publishers of the Lipsick Transactions." That the work was produced as a deliberate piece of Newtonian polemic is undeniable. Freind himself later admitted that "the grounds upon which I proceeded in my theory of chymistry, were the principles and method of reasoning, introduc'd by the incomparable Sir Isaac Newton; whose conclusions in philosophy are as demonstrative, as his discoveries are surprising."[54] In the work itself he was more careful, saying rather that "the way whereby chymistry might be examin'd and illustrated by mechanical, that is, by true principles, was first shewn us by Mr John Keill, a person who has very well deserv'd of the philosophical world."[55]

Freind's actual text was based on a series of nine postulates, borrowed directly from John Keill's 1708 theorems. With these as basis, the way was open to treat the whole of chemistry in scientific fashion (unlike that poor hero of yesteryear, Robert Boyle, who "has not so much laid a new foundation of chymistry, as he has thrown down the old"—so little did the new force-advocates acknowledge any debt to the old mechanical philosophies!). Of particular interest are Freind's explanations of why, though there is a weight increase on calcination, specific

---

[53] Freind (1709). Cf. Freind (1712a), *Chymical Lectures*. The dedication occurs only in the 1709 Latin version!

[54] Freind (1712a), *Chymical Lectures*, 173–174. This statement comes from his reply to the complaints voiced in "the Lipsick Transactions" or *Acta Eruditorum*.

[55] Freind (1709), 2–3. Translation quoted from Freind (1712a), *Chymical Lectures*, 4.

Atoms and Powers

gravity decreases, and of why different bodies are sublimed with varying difficulty.[56] His explanation of why aqua fortis dissolves gold but not silver, and aqua regia silver but not gold, which is "one of the most difficult and abstruse enquiries in chymistry," illustrates well the range and sophistication of his ideas. The account makes an interesting contrast with that put forward by Lemery only thirty-odd years earlier.

Freind's argument, given first in arithmetical terms, involved the relative sizes of the pores in silver and gold, the relative attractive forces of the two metals, their relative cohesions, etc. etc. He then went on:

> But to make the matter more general let us suppose the attraction of gold to that of silver, to be as a to b; and of silver to aqua fortis, as b to d; but that of aqua fortis to aqua regia, as d to e. Let f signifie the magnitude of the particles in aqua fortis, and r, those in aqua regia; c is the cohesion of gold, and g the cohesion of silver ... if $\overline{b-d} \times f$ exceeds g, then the silver will yield to the menstruum, whose particles are f, and less than the pores of the silver. And if $\overline{b-e} \times r$ is less than g, the silver will never dissolve in that menstruum, the particles of which are r, and the attractive force e ... But as yet we are not fully acquainted with the proportion there is betwixt the pores and cohesion of gold and silver ... tho' perhaps, when experiments are more accurately made, and examin'd by these mechanical principles, we may no longer remain in ignorance about them. At present, 'tis enough for our purpose, if from numbers and calculations, we can point out the way, which leads us to a solution of this phaenomenon.[57]

In the rest of the lectures, similar though less involved arguments were applied to other chemical phenomena, such as digestion and precipitation.

[56] Freind (1709), 15–16 and 34–37.
[57] *Ibid.*, 55 and 58–60. Translation quoted from Freind (1712a), *Chymical Lectures*, 95 and 100–102. For Lemery's earlier "mechanistic" explanation, see section 7.1. below.

The Immediate Impact of Newton's Ideas

The foregoing quotation shows what a sophisticated approach a reductionist such as Freind was able to advocate in the new, Newtonian, context. This approach stands in marked contrast to to the uncritical industry often displayed by later compilers of chemical affinity tables. Only a few minor investigators were at all quick to adopt Freind's ideas. The majority preferred to ignore his work. (Much later, when a Newtonian basis for chemistry came to be widely accepted in place of earlier Cartesian, "chymical," or empiric schemes, some of the most able chemical practitioners began to look back and hail his pioneer schemes.) Indeed it was one thing for Freind and his associates, "from numbers and calculations," to "point out the way" that chemistry should pursue, but quite another to carry out the sort of experiments on interparticle forces which these writers envisaged. As Mead's biographer put it in a remark which also throws light on his own period (the 1750's):

> In 1704 it was thought a very considerable thing to understand the system of Newton . . . and they who did so were willing the world should know it. It was the genius and universal fashion of the physicians at that time to introduce attraction into their art, as the modern ones have adopted electricity for the same purposes; though I apprehend the success hath not quite answered the endeavours of Cheyne, Keill, Freind and Mead.[58]

The sort of success the early Newtonians hoped for was not to come for over two centuries. They naturally could not know this. Yet they were uneasily aware that it was simpler to propound theoretical force laws, and to propagandize for a chemistry "reduced to the laws of Nature . . . which have never yet been apply'd to this subject," than to undertake the necessary experimental inquiries. It is therefore of some interest to look more closely at the practical work which took place under the auspices of the group.

[58] Mead (1755), *Authentic Memoirs*, 13.

Atoms and Powers

3.5. Early Experimental Physical Chemistry

The appropriate attempts to explore short-range force laws took place at the hands of three experimenters; Francis Hauksbee,[59] Brook Taylor,[60] and J. T. Desaguliers.[61] The first and last between them occupied the post of demonstrator to the Royal Society for over thirty years, including almost the whole of Newton's days as president. They undertook a host of experimental inquiries at his prompting. Brook Taylor—though influential in other capacities, especially the transmission of Newtonian ideas to France—was but a gifted dilettante in experimental work.

As soon as Newton became president, Hauksbee was employed with increasing frequency as demonstrator to the society, being elected F.R.S. in November 1705. Already in September of that year, Newton was writing to Hans Sloane to ask him "to get Mr Hauksbee to bring his air pump to my house, and then I can get some philosophical persons to see his experiments, who will otherwise be difficultly got together." [62] The relationship of these experiments to Newton's own on light and colors was widely recognized; and Hauksbee's work therefore aroused great interest. This interest may be seen in the correspondence of E. F. Geoffroy, who closely followed the "Newtonian" activities of the Royal Society, and himself produced the first published table of chemical affinities in 1719. As official correspondent for the French Académie Royale des Sciences, Geoffroy wrote to Hans Sloane in July 1709 that "we impatiently await details of the experiments Mr Hauksbee has made with his air-pump." In 1715 he was writing, in connection with a visit to London by his younger brother and the Comte de Monmort that "you will give them much pleasure if you are able to procure them the means of seeing Mr Hauksbee's experiments with his air-pump,

---

[59] (1660?–1717). See Guerlac (1964), "The ingenious Mr Hauksbee."
[60] (1685–1731). See Young (1793), *Life of Taylor.*
[61] (1683–1744). See Torlais (1937), *Desaguliers.* There is in the Library of Christ Church, Oxford, an excellent unpublished essay on Desaguliers by C. T. Newton.
[62] Nichols (1817–58), *Literary History,* IV, 59.

The Immediate Impact of Newton's Ideas

and those on colours reported in Mr Newton's book of opticks, if they may be seen."[63] The influence of Hauksbee's work is also apparent in the revisions George Cheyne made for the second edition of his *Natural Religion*.[64] However, the most important aspect of Hauksbee's experiments is the guide they provide to Newton's own preoccupations.

In 1704 Hauksbee began a series of experiments which developed into the first thorough investigation of static electricity and its effects. The work was described at length in his 1709 *Physico-Mechanical Experiments* in which he made plain the relevance of his work to the general theory of light and colors which "the most learned and incomparable Sir ISAAC NEWTON has invented and establish'd." And Newton's reaction to the information obtained via these electrical experiments may be discerned both in the last paragraph of the 1713 General Scholium and in the 1717 additions to the Queries.[65]

Through these alterations of Newton's, Hauksbee's work was to affect what later generations held to be Newtonian views of electricity, imponderable fluids and the ether. Of more importance to our present theme are the experiments that Hauksbee made on short-range interparticle forces. By the early eighteenth century, the phenomenon of capillary rise was well-known and under active investigation. As might be expected it was discussed by Newton in the last Query of the *Optice*. Instead of citing it as an example of the attractive forces at work in Nature, he at first used an explanation based on the repelling power of particles of air. The explanation came from a tract on the subject that Robert Hooke had published as long ago as 1661, and the argument of which Newton apparently still accepted in 1706. By 1717, when Query 31 appeared in English for the first time, the whole passage had vanished, to be replaced by four pages of experiments and commentary on capil-

---

[63] Quoted from letters cited in Cohen (1964), "Newton, Sloane and the Académie," 95 and 99.
[64] Compare Cheyne (1715), 89–90, with Cheyne (1705), 95–97.
[65] See Guerlac (1963, 1964, and 1967), where this matter is explored at length. Professor Guerlac's work has been freely drawn on for this account.

lary rise as an example of liquid-glass attractive forces! In this new section Hauksbee was mentioned instead of Hooke, and little research is needed to see how much its contents depended on the former's work.[66]

Hauksbee's own 1709 account of the experiments involved went on to consider "the solution of the phenomenon it self," saying that "it appears evident to me, that the principle we ought to have recourse to in this case, is no other than that of attraction." The statement was amplified in a passage which it is hard to believe Newton himself did not write, or at least approve for publication:

> If it should be thought that attraction is a word ... hard and unintelligible ... I can only say this, that 'tis plain fact that there is a power in Nature, by which the parts of matter do tend to each other; and that not only in the larger portions or systems of matter, but also the more minute and insensible corpuscles. And that the law which obtains in the former case ... is fully determin'd and settled ... But the law by which the smaller portions of matter tend to each other, is not so completely settled, but left yet for further discovery ... Only the fact it self is past dispute, and the discoveries made by that very great man, Sir ISAAC NEWTON, (the honour of our Nation and Royal Society) have set both these laws of attraction thus far, in a very clear light to all that will use their eyes to see them.[67]

In 1710, 1712, and 1713 the *Philosophical Transactions* contained further accounts of Hauksbee's work on capillary rise, and the attractive-power law associated with it. His experiments were of necessity crude. Yet they did demonstrate, to Newton's

---

[66] See the passage in Newton (1706), 340, beginning "Ex eo" and ending "explicavit D. Hookius." Contrast Newton (1718), 366–369. The abrupt reversal is characteristic of Newton's still unsettled thinking at this stage. It also exposes the hazards faced by early Newtonian commentators: e.g., consider the way in which the explanations offered in Harris (1710), *Lexicon Technicum,* article "air," so soon became redundant. The tract of Hooke's referred to is Hooke (1661).

[67] Hauksbee (1709), 157–158.

The Immediate Impact of Newton's Ideas

satisfaction if not to that of the French Académie, that there exist in Nature short-range attractive forces obeying laws other than that of the inverse square. Newton's own close interest may be seen from such entries in the *Journal Book* as that of January 1712, how it was desired that Hauksbee's paper on the ascent of oil of oranges "should be registered and published in the *Transactions;* as relating to the farther discovery of the nature of congruity or the agreement of the parts of matter." Again, on 26 June 1712, with "the President in the Chair," Hauksbee reported additional results on capillary rise which it was thought "might explain further the nature of attraction." And to quote from the 1717 version of Query 31—probably the only version that later generations consulted:

> By some experiments of this kind, (made by Mr. Hawksby) it has been found that the attraction . . . within the same quantity of attracting surface, is reciprocally as the distance between the glasses. And therefore where the distance is exceeding small, the attraction must be exceeding great . . . There are therefore agents in Nature able to make the particles of bodies stick together by very strong attractions. And it is the business of experimental philosophy to find them out.[68]

Hauksbee was prevented from further pursuing this "business of experimental philosophy" by his death in 1713, but some more work was done by his occasional collaborator, Brook Taylor, and by his successor, J. T. Desaguliers.

Brook Taylor was a comparative latecomer to the Newtonian circle we have been considering. He entered St. John's College, Cambridge, as a fellow commoner in 1701, and developed an interest in mathematics. From 1708 at the latest he was in regular correspondence with John Keill on mathematical matters. Elected F.R.S. on 20 March 1712, his first paper was presented

---

[68] Newton (1718), 368–369. And see *JB,* X, 358 and 411. The *Philosophical Transactions* for the years before Hauksbee's death carried a host of papers by him: thirty-three for the years 1704–1708 alone.

Atoms and Powers

on 25 June. It was "on the ascent of water between two glass planes." Taylor was secretary of the Royal Society from 1715 to 1718 (significantly he "was proposed with the general approbation of the Council by the President"),[69] and he visited Paris in 1716 before going on to Aix-la-Chapelle to convalesce on account of poor health. By his traveling and correspondence he played an important part in the communications between the Newtonian Royal Society and the Cartesian-inclined French Académie at this period, as we shall see in Chapter 4.

Only a one-page fragment of Taylor's paper appeared in the 1712 *Philosophical Transactions,* but even so its purpose was plain enough. As Taylor explained, "the following experiment seeming to be of use, in discovering the proportions of the attractions of fluids, I shall not forbear giving an account of it; tho' I have not here conveniencies to make it in so successful a manner, as I could wish." What are apparently further extracts from that same paper appeared in the *Philosophical Transactions* for 1715 as "An account of an experiment made by Dr. Brook Taylor assisted by Mr. Hawksbee, in order to discover the law of the magnetical attraction." This latter inquiry took its rise from a March 1712 proposal by Newton "that Dr. Halley and Mr. Hawksbee, should try the power of the great loadstone at several distances to find the true proportion of its decrease, which he [Newton] believed would be nearer the cubes than the squares." The final installment of the paper, only published in 1721, concluded that "at the distance of nine feet, the power alters faster, than as the cubes of the distances, whereas at the distances of one and two feet, the power alters nearly as their squares . . . From whence it seems to appear, that the power of magnetism does not alter according to any particular power of the distances, but decreases much faster in the greater distances, than it does in the near ones."[70]

The relevance of these magnetic experiments to Newton's

---

[69] *JB*, XI, 4.

[70] Taylor (1712), 538, (1715), 294, and (1721), 204–205; and *JB*, X, 375. See also the resulting alterations in the *Principia:* compare Newton (1687), 411 (Bk. III, prop. 6, corol. 4), with Newton (1713), 368 (Bk. III, prop. 6, corol. 5).

The Immediate Impact of Newton's Ideas

own thinking on the variety of forces in Nature, and to the modifications he made in the second edition of the *Principia,* is readily apparent. Indeed these attempts to establish the power law of magnetic attraction, and the variations in the attractive forces of fluids, must be seen not as a group of isolated and inconclusive experiments—which in one sense they were—but as conscious efforts by Newton's close followers to extend at his behest the quantitative approach he had used so powerfully in astronomy, and which he and they were convinced was equally applicable to short-range forces. In these efforts they were the more deliberate and determined because of Continental skepticism and hostility.

The *Journal Book* gives ample further evidence of Taylor's concern with attractive forces. For instance we read that

> upon [Hauksbee's] account of the attractive power of the magnet Mr Taylor said he had observed an attractive quality in two pieces of the same body, as wood or the like, and that if one were floating in water and the other were held at a small distance over it, the floating piece would follow it or endeavour to approach it. Mr Hawksbee was desired to consider of a way to try the like experiments as mentioned by Mr Taylor.

Perhaps because a way could not be found, or perhaps because the results were disappointing, the brief report published nine years later took a somewhat different line. Taylor said:

> I took several thin pieces of fir-board, and having hung them successively in a convenient manner to a nice pair of scales, I tried what weight was necessary, (over and above their own, after they had been well soak'd in water) to separate them at once from the surface of stagnating water . . . the weight in every trial being exactly proportional to the surface, I was encourag'd to think the experiment well made.[71]

---

[71] *JB*, X, 380, and Taylor (1721), 207–208: Taylor's suggestion was made on 3 April 1712, only two weeks after his election to the society—truly a promising disciple for Newton!

Atoms and Powers

An apparently trivial experiment—yet one powerful enough to awaken the interest of Guyton de Morveau half a century later, and to inspire his emulation at a time when chemists were rapidly being converted to Newtonian schemes.

Hauksbee died in 1713, and Taylor, plagued with ill-health, does not appear to have done serious experimental work after Hauksbee's death. Instead the task fell on the Royal Society's new demonstrator, J. T. Desaguliers. The son of a Huguenot refugee, Desaguliers, like Hauksbee, seems to have found experimenting for the Society an avenue of advance that supplemented and assisted his own career as popular lecturer.[72] Himself yet another member of the Christ Church group, Desaguliers gave decidedly more "popular" Newtonian lectures in Oxford than Keill ever did, following the latter's move to London. Abandoning Oxford for London himself in 1710, Desaguliers was well-received in Newtonian circles. He was proposed for membership of the Royal Society on 2 July 1713, the council duly recommending that "in consideration of his great usefulness . . . as curator and operator of experiments, he be excused from paying his admission money . . . and paying the weekly contributions." Not only that, but in due course he received regular payment (sometimes from Newton's own pocket) for his work.[73]

As with Hauksbee, so with Desaguliers, the mind behind the experimental work was that of Newton. And when Newton was absent, Desaguliers, true to his lecturing and technological concerns, turned not to experiments with magnets, thermometers and capillary rise, but to discussion of "fire engines," bridges, screws, and pulleys. However, until the closing years of his life, Newton was almost always present. As early as 18 February 1714, we find "that Mr Aguliers be desired to wait on the President to take his directions" as to some experiments with thermometers for great degrees of heat. Similar experiments re-

---

[72] This aspect of Desaguliers's activity is discussed in section 8.1. below.

[73] The quotation is from *Council Minutes of the Royal Society (Copy)*, II (1682–1727), p. 280, by permission of The Royal Society. For payments to Desaguliers, see for instance pp. 306, 309–310.

The Immediate Impact of Newton's Ideas

curred from time to time, including those that gave rise to Newton's "proof" of an ethereal medium from supposed heat conduction through a vacuum. And neatly combining such interests with a continuing inquiry into forces was an April 1714 "experiment of the proportional degrees of heat requisite for the fusion of several metallic bodies." [74]

As well as such obviously useful things as "experiments he had made in order to the explaining and verifying of the President's Treatise of Colours," Desaguliers was also engaged on continuing "experiments made by Dr Taylor and himself with needles touched with the loadstone," such experiments naturally earning further advice and suggestions from Newton. But the law of attraction and repulsion of the loadstone was difficult to discover. Rather different experiments, involving an iron wire floating on mercury, might "prove the mutual attraction of the parts of matter," but only in a disappointingly qualitative way. Again, electrical experiments "to show that a leaden ball of 4 ounces suspended by a thread may be drawn from the perpendicular by a rubb'd tube" were pretty enough, but not very helpful. Indeed with the laws governing short-range forces proving delightfully easy to theorize on, but distressingly hard to measure, much of the urgency seems to have gone from the work, especially after Leibniz' death. The one further reference—from Newton's extreme old age—is when Desaguliers in April 1725 "shewed an experiment on the prodigious force of cohesion in two leaden bullets joined by contact." [75]

The early Newtonian experimenters failed to quantify short-range forces, which remained obstinately complex, unlike the gravitational force. However, their work did serve as some sort of foundation and guide to more determined attempts by greater numbers of workers later in the century. Just as in this account the emphasis has been on chemically relevant aspects of the early work, so Chapter 7 will concentrate on later chemi-

---

[74] JB, X, 546 and 557. On Newton's "proof" of an ethereal medium, see Guerlac (1967), "Newton's optical ether."
[75] JB, XI, 16, 35, 126, 142; XII, 559.

cal quantification: yet in both cases the chemical work was but part of a wider endeavor to bring to triumphant fruition that "business of experimental philosophy" which Newton had begun. The early theoretical emphasis on short-range attractive forces was by no means solely motivated by scientific concerns, but can best be understood in the context of controversy with Leibniz (and of course with Cartesians—an aspect not stressed in this account). So also, the "nut-shell" theory, which so curiously left British Newtonians committed to infinite divisibility, yet uneasily aware of the atomistic nature of their master's thought, is best seen both against its immediate background of controversy and the broader context of theological unease over brute matter. The "nut-shell" theory was to condition chemical thought right down the century, often in Continental situations remote from the pressures that first led to its formulation as Newtonian orthodoxy. Finally, we should note here that Newton's early circle seems to have paid scant attention to the ethereal mechanisms he reintroduced late in life. This lack of attention must itself be understood in relation to their controversial preoccupations, even if it also reveals how far above their level were Newton's own insights and concerns.

The last major revisions to the *Opticks* were published in 1717. After that date Newton seems to have taken a steadily decreasing interest in controversy and experimental work (apart from some matters relating to the eventual third edition of the *Principia*). This is not surprising, for he was well past his appointed three score years and ten. From 1720 onward we may be said to be leaving the era of Newton and entering ever more deeply into that of "Newtonianism"—the curious and often inconsistent set of doctrines fathered on him by succeeding generations. The development of Newtonianism, its implications for chemistry, and its debts to the pioneer work of the early British circle must now be dealt with.

4
The Spread of Newtonianism: A Tradition Established

> Je suis venu a Leyde, consulter le docteur Borave sur ma santé, et Gravesande sur la philosophie de Neuton.
> *Voltaire in 1737*

4.1. Introduction

By the 1740's Newtonian doctrines had spread through most of Europe. These doctrines may be seen most clearly displayed in the textbooks—some elementary, some more advanced—of authors such as Desaguliers, Rowning, and Rutherforth in England, 'sGravesande and Musschenbroek in Holland, Algarotti in Italy, and Voltaire and Madame du Châtelet in France. The doctrines of these texts came to underlie the thinking of all those who approached chemistry from the reductionist standpoint of a physicalist natural philosophy. Thus later in the century, when this approach was finally establishing its intellectual dominance over both the recipe-collecting pharmaceutical and metallurgical (or "natural history") traditions, and the more intuitively chemical style favored by disciples of Stahl, it was to be in their Newtonian form that the beliefs of mechanistic natural philosophers became part of the base structure of chemical thought.

No account of the growth of Newtonianism can afford to neglect Newton's role, at least for the period into the early 1720's. An assessment of that role will be easier when the volumes of his *Correspondence* that cover the later years are finally published, but it is already quite apparent that Newton had a lively personal interest in the wide dissemination of his own ideas. His relations with the early British Newtonian circle, and with popularizers and translators such as Pemberton, 'sGravesande, and Coste, point to such a conclusion. And the surviving manuscripts of exchanges with such French scholars as Monmort, Varignon, and Fontenelle provide ample confirmation.[1]

By the time of the master's death, Newtonianism was flourishing. As it took root in such different cultural climates as those provided by the believing Protestants of the Dutch Republic or the skeptical *philosophes* of France, it not unnaturally assumed somewhat different forms. To analyze these forms, and to investigate the reasons (perhaps the aftermath of the struggle with Leibniz, perhaps the "Paracelsian" influence) why Newtonian doctrines made such slow progress among German chemists, is not within the scope of this work. We must rather be content with a rapid review of the growth of Newtonianism in France and Holland, and its relationship to British thought. More detailed study, and particularly investigation of the German unwillingness to accept Newton, would no doubt throw much further light on chemical developments later in the century.

4.2. *Newtonianism in France*

It is often assumed that French natural philosophers of the early eighteenth century were virtually unanimous in their opposition to Newtonianism and their defense of Cartesian orthodoxy, in their rejection of attractive and repulsive forces and their espousal of vortex mechanisms. Yet, as Brunet pointed out long ago,[2] the very eagerness to rebut Newtonian ideas which many French savants displayed, merely indicates the early and rapid infiltration of those same ideas. Recent investi-

---

[1] For letters by Monmort, see Kings College Cambridge, Newton Mss., no. 101. For Newton and Varignon, see Cambridge University Library, Mss. Add. 4007 ff. 586–589, 592–594, 612–617, and 625–654. For Fontenelle, see Bonno (1939), "Deux lettres à Newton."

[2] Brunet (1931), *Newton en France*, 7. Brunet's pioneer study of early French Newtonianism is now seriously dated, though not superceded.

The Spread of Newtonianism

gations³ have revealed how close was the contact of British and French men of science in the early eighteenth century. This contact was of course deeply affected by political events, notably the war of the Spanish Succession; but the surprising thing is the degree to which communication existed in this period, rather than the limitations placed upon it.

4.2.1. Early Contacts

E. F. Geoffroy,⁴ destined to become one of the most influential chemists in the Académie Royale des Sciences, spent most of the year 1698 in London. While there he was elected F.R.S. This election, and the Académie's subsequent appointment of Geoffroy as their official correspondent with the Royal Society, were to be of great importance. Also visiting London during the short interwar period round the turn of the century (the peace of Ryswick was signed in September 1697, and the war of the Spanish Succession began in May 1702) was Remond de Monmort.⁵ Subsequently Monmort became a close friend of Brook Taylor, with whom he conducted a vigorous controversy over "attraction." Monmort also carried on a friendly correspondence with Newton, some of which still survives in manuscript. Finally, as evidence of awakening French interest, we may note Newton's own election in February 1699 to the last of eight positions of *associé étranger* created as part of the reorganization of the Académie at this time.

The first edition of the *Opticks* was received in France with keen interest. At the Académie, Geoffroy gave a "presentation of Newton's *Opticks* . . . [which] extended over a period of ten months, from August 1706 to June 1707, and occupied a portion of each of ten meetings." Though what may possibly be the manuscript of the presentation has recently been located,

---

³ Notably Beer (1952), "The Royal Society and French science"; Jacquot (1953), "Sloane and French science," and (1954), *Les Exchanges Scientifiques*.

⁴ (1672–1731). See *Partington* (1961–64), III, 49–55, and Cohen (1964), "Newton, Sloane and the Académie," which has been freely drawn on in what follows.

⁵ (1678–1719). See Fontenelle (1719), "Eloge de Monmort."

85

we have no knowledge of the academicians' reaction.[6] However, in the case of the 1706 Latin edition, for which of course translation was unnecessary, we do have the evidence of summary reviews in both the *Journal des Savants* and the Jesuit-controlled *Journal de Trévoux*.

As was its fashion, the *Journal des Savants* contented itself with a competent but noncommittal summary of the book's contents and arguments (though it also reproduced a long quotation from Mariotte's 1681 objections!). Of the Queries it merely noted that "[Newton] has increased the number of questions at the end of the third book."[7] The *Journal de Trévoux* was both more perceptive and more provocative. Its review began by asserting that while Newton's work on light had made the history of Nature more complete, it had made the research into causes more difficult. A noncommittal summary of the book followed once more, but this time it concluded by contrasting "our physicists, who everywhere consider mechanisms," with Newton, "whose grand principle is attraction." The difference was then driven home in a passage which began, "There is no chemical operation which is not to him a good proof of the efficacy of attraction," went on to discuss Newton's observations on salt of tartar in the 1706 Queries, and ended by contrasting his explanations of light transmission with Cartesian ones.[8] This interesting review thus makes it clear that, as early as February 1709, there were in France those who saw plainly not only the opposition between the rival theories of light but also the chemical implications of Newton's work.

It was of course at about this time that Francis Hauksbee was busy with his experiments. These experiments provided excellent material for a system based on attractive and repulsive forces, and were incorporated by Newton into the 1717 Queries, along with a revised explanation of capillary rise. Hauksbee's

---

[6] Cohen (1964), "Newton, Sloane and the Académie," 107.

[7] *Journal des Savants* (1707), 139 (cited below as *JdS*). On the journal's history and connection with England, see respectively Morgan (1929) and Harpe (1937–41).

[8] *Journal de Trévoux* (1709), 185–201 (cited below as *JdT*).

work was published in the *Philosophical Transactions* from 1706 on, and we have already noted the keen interest his work aroused in Paris. The whole question of French and British work on capillary rise over the period 1704–1724 would repay examination for the light it throws on early Newtonianism. Here we may just note that Geoffroy's 1710 receipt of a copy of Hauksbee's *Physico-Mechanical Experiments* must have further stimulated that curiosity about Newton's ideas on short-range forces already apparent in France.[9]

One additional piece of evidence from this early period deserves mention. It shows so well both the interest of the French and their awareness of the chemical implications of Newtonianism. Such an important polemical piece as John Freind's 1709 *Praelectiones Chymicae* had quickly earned that standard device for Continental diffusion, an Amsterdam reprint. This reprint was soon reviewed in the French periodicals. The *Journal des Savants* was again first and least perceptive, contenting itself with a brief summary of the book, beginning: "The design of the author, as he explains it to the illustrious Mr. Newton, to whom he dedicates his work, is to explain the operations of chemistry by the attractive virtue." The *Journal de Trévoux* devoted more space to Freind, reviewing his work favorably, and noting that in his view attraction was "the general principle that Nature employs." In neither journal was there any attempt to attack, confute, or mock at Freind's work—a situation which contrasts strongly with the treatment accorded his book by the *Acta Eruditorum*.[10] It is thus apparent that in roughly the opening decade of the century, though at war with England, the French were keenly interested in Newton's work, and well aware of its chemical implications, towards

---

[9] Cohen (1964), "Newton, Sloane and the Académie," 97. For French contributions on capillary rise, see e.g., *Histoire de l'Académie Royale des Sciences* (1705), 21–25 (cited below as *HAdS*); *Memoires de l'Académie Royale des Sciences* (1705), 241–254 (cited below as *MAdS*); *HAdS* (1722), 35–36, and (1724), 1–14; *MAdS* (1724), 94–107.

[10] *JdS* (1711), 10; *JdT* (1712), 1780–1786. See section 3.4.1. above for the Hanoverian reaction to Freind.

which they were not unfavorably disposed. At this stage there was little sign of that hostility towards the "occult" nature of attraction which was so to complicate the French scene over the next thirty years.

4.2.2. Visits, Controversy, and Newtonian Chemistry

While these earlier discussions of Newton's optical work and its chemical implications were marked by their calm and sympathetic nature, ominous storm clouds were gathering. In the years after 1710, the Newton-Leibniz dispute led to an ever-widening rift between British and Continental (or at least German, Swiss, and French) mathematicians and natural philosophers. The events and controversies we have already touched on in Chapter 3 were to lead between 1710 and 1720 to a polarization of attitudes throughout Europe, and were to create bitterness and rivalry between the emerging groups of British and Dutch Newtonians and their French and German protagonists.

It is against this changing background that we must see the 1713 second edition of the *Principia*. This edition was eagerly awaited on the Continent. Evidence of Newton's own desire for its wide distribution may be seen in his sending both a personal copy to Fontenelle, and an official copy to the French Académie.[11] The long anti-Cartesian preface by Cotes, and the concluding General Scholium which began "The hypothesis of vortices is pressed with many difficulties," can only have served to intensify Cartesian opposition, and to develop French sympathy for Leibniz. The extent of Newton's own involvement in the subsequent controversies is only now coming to light. This involvement does not concern us directly. We must rather attend to French discussion of Newtonianism, which now became more urgent.

---

[11] Edleston (1850), *Newton-Cotes Correspondence*, 235. A further indication of how rapidly the various editions of Newton's books were transmitted abroad, and of Newton's own involvement in the process, is apparent in a draft of a 17 November 1718 letter to Varignon, telling how "a few weeks ago I [Newton] sent you a copy of a new edition of my English Opticks": Cambridge University Library, Ms. Add. 4007 f. 587.

The Spread of Newtonianism

The formal conclusion of the war of the Spanish Succession in May 1713 allowed a renewed flow of scientific travelers between England and France. The visit to London in 1715 of Remond de Monmort and the younger Geoffroy has already been mentioned. During this visit, Monmort was made an F.R.S., saw Newton's optical experiments demonstrated, and became a firm friend of Brook Taylor, who at the time was secretary of the Royal Society. A return visit by Taylor to Monmort in Paris soon followed. Records of some of their correspondence in this period still survive.

Typical of the close relations of the British Newtonians, and of French concern, is a letter of Taylor's to John Keill, dated 17 July 1717. This includes the request "to put Mr. Innys in mind to send me ... two copies of Sir Is: Newton's *Opticks*, as soon as it is out, one bound, & another in sheets, which I must send to Mr. Monmort." More revealing as to Anglo-French controversy about Newtonianism is Monmort's letter to Taylor of 31 March 1716. This refers to "your objections against our fundamental principles and our method of pursuing natural philosophy" and goes on to confess sadly that "we are even more divided over natural philosophy than our theologians are between themselves." The edge provided to the quarrel by the calculus dispute may be seen in Monmort's later note about how "I cannot refrain from combatting the opinion which you hold that the public received from Mr. Newton, and not from Messrs. Leibniz and Bernoulli, the new calculus," and how "some Englishmen seem to bear a grudge against those who have first discovered and published these new methods." [12]

Taylor's biography merely observes that "the correspondence which he had some time engaged in with learned foreigners, and more especially the French, on the various topics of the new philosophy, had given him a general reputation abroad." Fontenelle's 1719 eulogy on Monmort is more revealing. The importance of the friendship, correspondence, and visits, as

[12] Young (1793), *Life of Taylor*, 85, and Brewster (1855), *Memoirs of Newton*, II, 513, 515.

well as the prevailing mood of the Académie in 1719, is well brought out in the statement that "Mr. Monmort had great quarrels on this subject [attraction] with Mr. Taylor, his bosom friend. Mr. Monmort composed a long dissertation with the utmost care, hoping to drag attraction down into the dust, from whence it would be unable to rise. Mr. Taylor rapidly replied. Certainly if one takes what he says at face value, there is only impulsion, and if one does not take it at face value, there is attraction, and whatever else one wants." [13] Bearing these things in mind, we can perhaps appreciate how charged was the atmosphere into which E. F. Geoffroy was to introduce his seminally important 1718 "Table of the different relationships observed in chemistry between different substances" [14] (see Fig. VI).

Though couched in language of studied neutrality, Geoffroy's work was immediately suspected of Newtonian inspiration and implication. After all, it was in the 1706 *Optice* and again in the 1717 English edition that Newton had provided just such a list of relative attractions as was to form the basis of Geoffroy's table. Geoffroy's own earlier visit to London, the favorable reception he there enjoyed, and his resulting friendship and regular correspondence with Hans Sloane were all calculated to make other academicians suspicious for his Cartesian orthodoxy in this period of hostility. To what extent Geoffroy was a secret Newtonian we do not know, though evidence exists of his sympathy with Newton's viewpoint.[15] He certainly made no public attempt to disavow the implications of his table.

Fontenelle's eulogy provides ample confirmation of the suspicion with which the table was greeted: "These affinities upset some, who believed they were only disguised, and therefore more dangerous, versions of attraction (for clever men had already succeeded in giving attraction seductive forms)." In his earlier commentary, published along with the table itself

---

[13] Young (1793), 22, and Fontenelle (1719), 91.
[14] Geoffroy (1718).
[15] Cohen (1964), "Newton, Sloane and the Académie," 100–102.

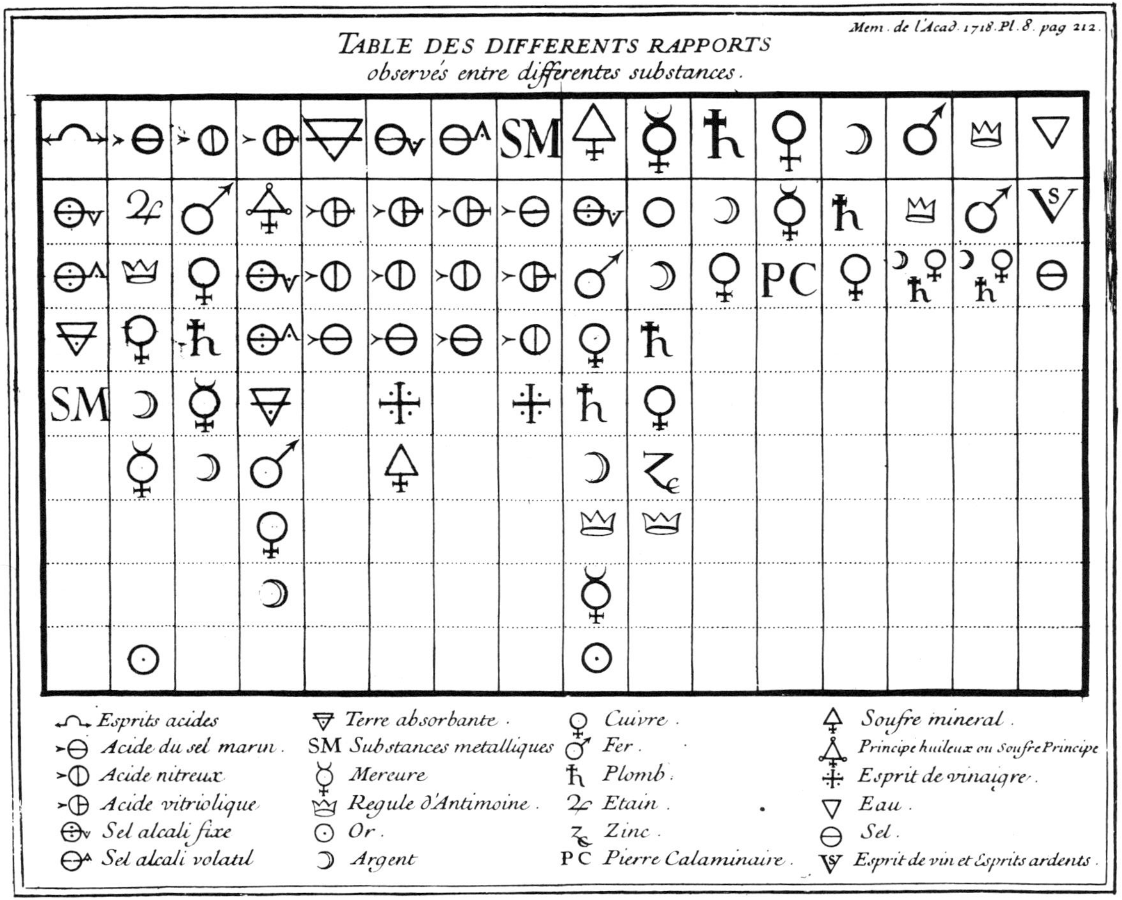

VI. Geoffroy's affinity table. From Geoffroy (1718), plate VIII.

in 1719, Fontenelle contented himself with saying that preferential reaction was not easy to explain and that "it is here that sympathies and attractions would fit in well, if there were such things. But in leaving for unknown . . . [the causes of preferential reaction we hold instead] to definite facts." Along with this brief anti-Newtonian polemic, he was also at pains to say that "the more chemistry is perfected, the more Mr. Geoffroy's table will also be perfected," and that "if natural philosophy cannot hope to arrive at the certitude of mathematics, even so it cannot do better than imitate the latter's order. A chemical table by itself is a pleasing sight, as would be a numerical table arranged according to known relationships or properties." [16]

In his dual emphasis on "definite facts" and a numerical table that would "imitate the order" of mathematics, Fontenelle was prefiguring much later French thought. When Newtonianism finally gained ascendancy in France, it was to be the mathematical-positivistic rather than the speculative-experimental version of that creed which appealed to such men as Laplace, Lagrange, and Lavoisier. The positivistic, if not the mathematical, form taken by so much French Newtonianism is readily understood if we appreciate the far more hostile climate that France provided for any explanation of physical forces based on religious principles. Philosophers anxious to attack Cartesianism (and the religious orthodoxy with which it was associated by the early eighteenth century) would scarcely wish to emphasize the religious basis of the new science they preferred!

Even in the "high period" of French opposition to Newtonianism—say the twenty years following 1712—one of the most interesting things is the continuing awareness of, interest in, and sympathy towards Newton's ideas on light, color, and chemistry. Though the attitude of Fontenelle and his collaborators may have prevented the acceptance of Newtonianism by the educated lay public, it is clear that many scholars were well disposed toward Newtonian ideas by the early 1720's. It is also clear from Geoffroy's uncommitted article, and Fontenelle's

[16] Fontenelle (1731), 99–100, and (1718), 36–37.

Cartesian reaction to it, how the French were aware that Newtonianism was a doctrine embracing not merely celestial mechanics and optics but the whole of chemistry.

Pierre Coste's French translation of the *Opticks* was published in Amsterdam in 1720, a much improved second edition appearing in Paris in 1722. The translator's preface to this second edition acknowledged the help of J. T. Desaguliers "who has had the goodness carefully to review my manuscript." What it did not mention was Desaguliers's position as trusted servant of Newton, or Newton's own active interest and cooperation in the venture—a fact, as so often, not apparent from the work itself but only from surviving correspondence. Newton's desire to have his work known may be seen again in his present of a copy to the Académie (though he was careful not to disclose his part in the translation).[17]

In the preface to the work, Coste was at pains to emphasize that the more difficult of Newton's optical experiments had been successfully repeated in Paris before an audience including such notables as Varignon and Fontenelle. He was also careful to point out that the Queries had "solid foundations," that "one will find there what the author thinks on the most important parts of physics," and that "one can also see by what he says of attraction that this principle is not in the least what the schoolmen have named an occult quality." [18] All these bold claims raised no objection. The *Journal des Savants,* in its usual noncommittal review, even pointed out that "the work finishes with several questions which serve to confirm the earlier parts." The *Journal de Trévoux,* though careful not to side openly with Newton, could not resist a panegyric on his qualities.[19]

As a final example of the sympathetic hearing accorded in this period to Newton's views on matter, we may mention the

---

[17] Coste (1722), sig. a ii^r. See Kings College Library, Cambridge, Newton Mss. no. 94, for a letter from Coste to Newton. As Dr. D. T. Whiteside has pointed out to the present author, Newton's involvement in the French edition of the *Opticks* was first noted by Brewster (1855), *Memoirs of Newton,* II, 501.

[18] Coste (1722), sigs. e ii^r & e iii^{v,r}.

[19] *JdS* (1720), 545 (review of first French edition), and *JdT* (1723), 1428–1450.

reception given the anonymous *Nouveau Cours de Chymie Suivant les Principes de Newton & de Stahl* of 1723. This work was in large measure a straight translation, without acknowledgement, of the writings of Keill and Freind, especially of the latter.[20] The *Journal des Savants* declared that "in this new work chemistry offers itself under a different face; in it one sees this mysterious science divest itself of obscurity, in favor of the light of physical argument. In it mathematics brings its rules to the philosophy, and experience confirms the reasoning." The *Journal de Trévoux* showed still greater enthusiasm, producing a review which ran to fifty pages, and quoting long sections of the book verbatim. Two points of particular interest in this review are its dismissal of Nicholas Lemery and its careful discussion of attraction and affinity.

Lemery's Cartesian-inspired *Cours de Chymie* had been available for more than half a century, and at this time was probably still the most popular chemical textbook in Europe. The review compared it unfavorably with the *Nouveau Cours* saying: "[Lemery] never loses sight of the acid-points which break on hitting the alkali; this is the source of his explanations for all the phenomena of chemistry. A natural philosopher who wishes only for clear principles, drawn from the laws of motion, will not be able to content himself with this." The discussion of attraction made it plain that this term was not exceptionable if understood correctly, that is positivistically, a point Newton himself had been at pains to emphasize in his more public statements. Thus:

> In the *Opticks* Mr. Newton discussed chemical precipitations. This great man made it clear that one could not hope

---

[20] For example, compare the passage on Boyle, Senac (1723), I, 2–3, with that in Freind (1709), *Praelectiones Chymicae*, 2–3. On Senac, see *Partington* (1961–64), III, 58–59. There is some doubt as to whether Senac was the author of the *Nouveau Cours:* see the article on Senac in Michaud's *Biographie Universelle*, and the 1768 query from P. J. Macquer to Torbern Bergman in Carlid (1965), *Bergman's Correspondence*, 230. Partington's suggestion that the *Nouveau Cours* derived from Geoffroy's lectures is based on an incorrect reference. Whoever the author, the work (which enjoyed a second edition in 1737) testifies to early French interest in Newtonianism.

to give them a reasonable explanation without supposing a cause which unites certain bodies, some more strongly, others less so. He called this unknown cause *attraction*. It has since been disguised under the name of affinity, which is no clearer a term than attraction ... [When the author] says that it is magnetism, or attraction, that produces certain effects, he does not pretend that this is a physical reason; he only wishes to give a sure rule to judge what happens in an infinitude of reactions, ignoring the reasons and illuminating the facts.[21]

The application of Newton's ideas on short-range forces to the whole range of chemical phenomena was thus widely appreciated, if not actively pursued, in France. With the possible exception of Geoffroy's work, there seem to have been no French attempts to develop a Newtonian chemistry at this early stage. However the interest in and appreciation of Newton's ideas, apparent right through the first quarter of the eighteenth century, was to have important consequences when Newtonianism was finally accepted by educated lay opinion. This earlier knowledge of the work of Newton and his first-generation disciples such as Keill, Freind, Taylor and Hauksbee, was to provide a rich store for such later workers as Macquer, Guyton, and Berthollet to draw on. The crucial link between the earlier awareness and the later flowering of French Newtonian chemistry was to lie in the writings and patronage, especially the patronage, of Georges-Louis Leclerc, Comte de Buffon. His work will be dealt with in Chapter 5. Here it is only necessary to consider briefly the final establishment of Newtonianism in France.

4.2.3. Newtonianism Established

The discussion so far has been concerned with reaction in France to Newton's ideas, especially his chemical ideas. This discussion has shown that neither ignorance of, nor hostility to,

---

[21] *JdS* (1724), 80; *JdT* (1723), 2020, and (1724), 198–200. Lemery's widely influential textbook was originally published in 1675, and reached its tenth edition by 1713. For Lemery's ideas on reaction mechanisms, see section 7.1. below.

these ideas was as widespread as has been imagined. Yet Cartesian enthusiasms, sympathy for the Leibnizians, and suspicion of English motives all remained. Before 1730 no major figure in the French Académie was prepared publicly and unequivocally to support the Newtonian position against such a powerful combination of adversaries. Such a figure was soon to appear, however, in the person of P. L. M. de Maupertuis.[22] The story of his serious, academic support of Newtonianism is closely interwoven with that of Voltaire's activity on the cruder, popularizing level. While no sudden change took place, their combined activity was sufficient to tip the balance of educated opinion. Thus by 1743, D'Alembert was referring to the Cartesians as "a sect that in truth is much weakened today," and by 1758 this had become "a sect that in truth hardly exists today."[23]

Maupertuis and Voltaire both visited London, independently, in the late 1720's. Maupertuis's reception in British scientific circles, his friendship with such prominent disciples of Newton as John Keill, and his election to the Royal Society were all doubtless of importance in encouraging his adoption of a Newtonian position. In 1732, soon after his return to France from a variety of travels, he published two important works which made apparent his sympathies. The one was his *Discours sur les Differentes Figures des Astres* which included "a summary of the Cartesian and Newtonian systems," from which Maupertuis's preference for the latter was plain. Of greater interest for this present study is his other work, a paper "Sur les loix de l'attraction" in the *Mémoires* of the Académie.

---

[22] (1698–1759). Brunet (1929 a and b) provides careful and thorough studies of his life and scientific thought.

[23] Quoted from Guerlac (1965), "Divergent loyalties to Newton," 318. Consider also the revealing remarks of Benjamin Wilson in his 1761 edition of *Robins' Mathematical Tracts:* "And indeed Sir Isaac Newton's fame seems at present to have surmounted all opposition. The philosophers of a neighbouring nation [France] acknowledge his merit. Though they had for years quite overlooked his book of *Opticks;* yet they now speak of it in the highest terms. And they have at length adopted attraction under the name of universal gravitation; which they had long rejected as an occult quality, on account its cause is not discovered." Wilson (1761), I, xiv.

The Spread of Newtonianism

The paper was in the form of a commentary on sections 12 and 13 of Book I of the *Principia*. It provided a sophisticated mathematical discussion of the possible variations in attractive force between "an atom or very small body placed externally on the axis of a spherical mass" and the mass itself. Before beginning this discussion, Maupertuis was careful to point out, in the approved manner, that "he did not examine whether attraction is repugnant to or accords with sound philosophy, but only treated it geometrically." Nonetheless his attitude was plain, not least in the closing sections where attractive forces decreasing more rapidly than $1/r^2$ were discussed. Keill's laws "which form a small treatise of thirty propositions" were mentioned, as also the fact that "Keill and several English philosophers, following Newton, believe they have found in bodies . . . another species of attraction . . . capable of producing precipitation . . . and an infinity of other phenomena which have previously been attributed to more respectable-sounding adhesions and affinities. And Freind has even produced a chemistry completely deduced from this new principle." [24]

Once again we see how common was the French appreciation of the chemical possibilities of Newton's work, and how well known the writings of Keill and Freind. The concluding section of Maupertuis's article is also of great importance because of its treatment of a problem which, in a different form, was to intrigue and engage Buffon and the French Newtonian chemists who followed his lead. Maupertuis mentions how, after following the rest of his program, "one can solve this curious problem: *to discover what law the attraction of a given substance follows.*" In its reverse form of discovering the shape of a particle whose law of attraction was known, it was this Newtonian problem that Buffon and his followers were to see as the key to chemical understanding.

Fontenelle's commentary on the paper, published in the Académie's *Histoire,* was restrained. He contented himself with saying that "[attraction] is only a name given to an unknown cause, of which the effects are felt everywhere." Greater interest

[24] Maupertuis (1732), 343–344 and 361–362.

attaches to his discussion of the possible law followed by "the primitive attractive force." Whereas Newton was apparently satisfied, at least in his public utterances, with a variety of attractive and repulsive forces, Fontenelle, having grudgingly accepted the utility of "forces," was already seeking a more primitive or fundamental law in a manner typical of Buffon and other systematizers. The central role of chemistry in any discussion of the Newtonian system is again apparent, for

> however good the theory of attraction is for geometrical use, it is easy to see that its application to Nature will always be too difficult—above all in the choice of the true primitive law of attraction. That of the inverse square of distance succeeds in physical astronomy [as Newton has shown] . . . Other English *savants* have with great reason believed that it ought also to be extended to terrestrial phenomena, principally to chemistry, which gives an idea of attraction incomparably more striking than all celestial phenomena. With what impetuosity certain acids try to penetrate their proper alkalis! What a commotion in the reaction vessel! But all this is too violent for the system it appears to favor . . . It will be the task of future natural philosophers to show that a certain supposed primitive attraction satisfies all the phenomena, both terrestrial and celestial. These natural philosophers need not fear lack of employment: geometers will lack it sooner.[25]

The endeavor to quantify the force law of chemical attraction, in the way that Fontenelle here half-ironically envisaged, was to become a major obsession of chemists, especially French chemists, in the latter part of the century. Against the twin backgrounds of a Newtonian astronomy gaining ever greater prestige through the successes of such men as Maupertuis, Clairaut, Lagrange, and Laplace, and a wide awareness of the early attempts to provide a theoretical statement of the power laws of chemistry, it is understandable how French attempts

[25] Fontenelle (1732), 116–117.

to quantify and thus control this science were to be firmly based on the Newtonian dream of measuring short-range interparticle forces.

It is thus somewhat surprising to find that, as late as 1733, Voltaire [26] was declaiming, with typical exaggeration, how "a Frenchman who arrives in London, will find philosophy, like everything else, very much chang'd there . . . At Paris . . . the several operations of your chymistry are perform'd by acids, alkalies and subtile matter; but attraction prevails even in chymistry among the English." As is apparent, this statement woefully underestimates previous French interest in Newton's work, but it is at least as representative of the deep divide between the Newtonian and Cartesian world-views as Voltaire's far more quoted statement about the plenum and the vacuum. The story of Maupertuis's influence on Voltaire ("me voicy neutonien de votre façon") is well known—Maupertuis even read the manuscript of the *Lettres Anglais* before publication.[27] However while the *Lettres* were undoubtedly of great propaganda value, it was to be Voltaire's 1738 *Elémens de la Philosophie de Newton* that was of more immediate utility in the creation of a body of Newtonian lay knowledge.

The reader of the *Elémens* would be made aware of the internal structure of matter, the existence of indivisible atoms, and the variety of forces in Nature. However he would not meet either the ether or the more speculative parts of the Queries—subjects too embarrassing for anyone wishing to show the differences between Newton's hard, practical approach and

---

[26] (1694–1778). Voltaire enjoys much the same sort of attention from historians of the Enlightenment as Newton does from historians of science. The new 107-volume edition of his correspondence (Besterman [1953–65]) and the existence of the special journal *Studies on Voltaire and the Eighteenth Century* (ed. T. Besterman, Geneva, 1955–   ) are perhaps dire harbingers for Newtonian studies. By 1968 no less than sixty-eight volumes of the journal had been published, volumes which contain many articles that deserve to be better known among historians of science (e.g., Crosland [1963], Kiernan [1968]) as well as studies of every facet of the Enlightenment.

[27] Quotations from Voltaire (1733b), *Letters*, 109–111, and letter no. 516 in Besterman (1953–65). Other letters of the period that deal with Voltaire, Maupertuis, and Newtonianism are Besterman nos. 515–517, 520, 523, 528, and 532.

Descartes's speculative, hypothetical science. Still, he would be conditioned toward that view of matter, its forces and its properties, soon to be taken as self-evident by natural philosophers. And if instead of Voltaire he were to turn for elementary instruction to Madame du Châtelet, mistress of Voltaire, equally ardent Newtonian and author of the 1740 *Institutions de Physique,* there too the message would be the same. There also he would be clearly informed that "Newtonians, who make attraction an inseparable property of matter, wish it to reign everywhere; but when by its means they wish to explain cohesion, chemical effects, the phenomena of light etc., they are obliged to suppose other laws than that which directs the course of the stars." [28]

The Newtonian texts of Voltaire and Madame du Châtelet were also supplemented by the translation of Dutch and British works. We may note, for instance, Voltaire's 1737 visit to Leyden to meet W. J. 'sGravesande, and the latter's sustaining and directing of his Newtonian convictions. We may also note the appearance of Musschenbroek's weighty 1736 Dutch Newtonian text (*Beginsels der Naturkunde*) in French translation in 1739. And the potential market in France, in 1734, was apparently already such that in England J. T. Desaguliers felt it necessary to prefix to his *Course of Experimental Philosophy* the statement that "whereas some booksellers have declared, that as soon as my course comes out, they will get it translated into French . . . I intend to translate it my self, having already done more than half." [29] On the level of original works rather than textbooks, Buffon's 1735 translation of Stephen Hales's *Vegetable Staticks* marked a further important step in the diffusion of Newtonian

---

[28] See Voltaire (1738b), *Elements of Newton,* 30–31, and chap. 10. The extract from the 1740 *Institutions* is quoted in Metzger (1930), *Newton, Stahl, Boerhaave,* 41. For a lighthearted but informative study which includes much incidental information on French Newtonianism in the 1730's and 40's, see Mitford (1957), *Voltaire in Love.*

[29] On the importance to France of Dutch Newtonianism, there is an excellent monograph: Brunet (1926), *Les Physiciens Hollandais.* For Voltaire and 'sGravesande, see Besterman (1953–65), nos. 1204, 1222, 1248, and 1253. The quotation is from Desaguliers (1734–44), I, sig. c 4ʳ.

ideas. By the 1740's, therefore, Newtonianism, if not undisputed, was yet well established in France.

4.3. Newtonianism in Holland

While the story of the reception of Newtonianism in France is one of caution, suspicion, and sometimes overt hostility, the tale of its reception in Holland is essentially one of acceptance and development. This openness toward British ideas was undoubtedly aided by the existence of similar theological traditions, and the political and military alliances between the two countries in this period. (Easier travel and communication was one obvious consequence.) Characteristically, one of the three major Dutch Newtonians of the early eighteenth century was first brought into direct contact with things English for political and diplomatic, rather than scientific, reasons.

4.3.1. 'sGravesande

W. J. 'sGravesande [30] was educated at Leyden University, where Herman Boerhaave was already one of the professors. In 1714 'sGravesande was chosen to be in the delegation sent to England by the Dutch States-General on the accession of George I. Through former Leyden students then in London, he quickly made contact with the Royal Society, to which he was duly elected. We soon find him being "entertained" by the experiments of Desaguliers, and he apparently became a personal friend of Newton. Indicative of the close relations that prevailed is a Royal Society minute of 2 February 1716 that "Mr. 'sGravesande told the President he was going into Holland the beginning of the next week, and desired to know if he could do the Society any service there." [31] Characteristically Newton, in

---

[30] (1688–1742). See Marchand (1758–59), *Dictionnaire Historique*, II, 214–242, for a highly informative biography.

[31] Quotation from p. 100 of *Journal Book of the Royal Society (Copy)*, XI, 1714–1720 (cited below as *JB*). As early as 24 March 1715 it was recorded (p. 57) that "Mr Desaguliers shewed several experiments to entertain some foreign gentlemen, viz. Mr Gravesend, the Secretary . . . to the present Dutch Embassy . . . who were present at the meeting of the Society." Already four weeks earlier

his role as zealous and money-conscious administrator, took advantage of the offer by desiring the collection of some debts!

On his return to Holland, 'sGravesande was appointed to a chair of mathematics at Leyden University, and in 1720 the first of the two volumes of his *Physices Elementa Mathematica* appeared. Revealingly, a copy of the first volume was already at the Royal Society by 22 October 1719, and Desaguliers's translation was apparently made by 7 January 1720—portions being read to the society on the 28th, and the printed English version appearing at the same time. The *Physices* was the earliest systematic textbook of natural philosophy to be of deliberate Newtonian persuasion. Being written in Latin, it was of course freely accessible to educated readers throughout Europe—a critical review in the *Journal de Trévoux* for May 1721 bears witness to its speedy diffusion. A rival English translation quickly appeared,[32] and the wide popularity of the work may be gauged from the appearance of a fifth English and third Latin edition by the early 1740's.

As the full title (*Mathematical Elements of Natural Philosophy, Confirmed by Experiments, or An Introduction to Sir Isaac Newton's Philosophy*) made plain, 'sGravesande's was a thoroughly "Newtonian" text. The discussion of the infinite divisibility of matter, and the actual small particles of which it is composed, reproduced the first of John Keill's "theorems" on this subject. Potential infinite divisibility and actual hard particles were apparently simple to reconcile, but Newton's (re-

---

(p. 49), "Mr Burnet proposed Mr Gravesend . . . to be chosen a Fellow of the Society," so good were the Leyden-London contacts. For a 1721 letter from 'sGravesande to Newton, see Kings College Cambridge, Newton Mss. no. 94. For later communications from 'sGravesande to the Royal Society, see *JB*, XII, 265, 609; XIII, 5. 'sGravesande was actually in correspondence with Newton as early as 1714: see Cambridge University Library Ms. Add. 3968 ff. 594–595.

[32] See 'sGravesande (1720 a and b). The "authorized" translation was to be that by Desaguliers, but so great was the interest that a rival bookseller threatened to scoop the market with his own version, purportedly "revis'd and corrected by John Keill." Desaguliers's anguish, and side lights on the book trade, may be gleaned from his unpublished letters to Keill in Cambridge University Library, Lucasian Mss., packet no. 3.

dundant) assumption of different sizes and shapes among these fundamental particles was questioned in the statement that "I am entirely ignorant whether all bodies consist of equal and like particles." Apart from this reservation, the Newtonian picture of a hierarchical internal structure to matter was faithfully reproduced.

Newton's ideas on the role of short-range attractive and repulsive forces were also carefully set forth. First 'sGravesande made it clear that "by the word attraction I understand *any force by which two bodies tend towards each other;* tho' that, perhaps, be done by impulse." He then went on to discuss the variation of the law of attraction with distance in a way which reveals how much this subject exercised early Newtonian commentators. In an unpublished discussion Newton himself had gone so far as to picture how "the particles of bodies have certain spheres of activity within which they attract or shun one another." And in 1735 John Rowning, one of the second generation British Newtonians, was to develop the same point, by speculating whether "each particle of a fluid must be surrounded with three spheres of attraction and repulsion one within another: the innermost of which is a sphere of repulsion, which keeps them from approaching into contact; the next, a sphere of attraction . . . by which the particles are disposed to run together into drops; the outermost of all, a sphere of repulsion." Though well connected among the early Newtonians, 'sGravesande was much more cautious in his own discussion, merely saying that "attraction is subject to these laws, that in the very point of contact of the particles it should be very great, and should suddenly decrease; so that at the least distance that can be perceived by the senses, it should act no more; nay, at a greater distance should change itself into a repelling force, by which the particles mutually fly off from one another."

'sGravesande's own more orthodox treatment explained how, "by the help of this law, a great many phaenomena are easily explain'd: This attractive and repelling force is prov'd by innumerable chymical experiments." Among the examples given of this "attractive and repelling force" were capillary rise, the ascent of water between glass planes and a variety of other examples drawn from Hauksbee's work and the Queries in *Opticks*. More directly chemical examples of attractive force included the action of mercury on gold or copper, while that "between quicksilver and iron" illustrated the repelling force. The solution and crystallization of salt, following the Queries, exemplified both attraction and repulsion.[33]

'sGravesande's book thus demonstrates how well he spent his time in London, and how faithfully he mastered the Newtonian approach—not only on the internal structure of matter and short-range forces, but also on all those questions of Newtonian mechanics and astronomy which lie beyond the scope of this inquiry. Again, his defense of attractive and repulsive forces against the charge that they were occult qualities was taken almost word for word from Newton—indeed his only significant deviations from Newton's 1720 position lay in his failure to discuss the ether, and in his extended treatment of fire.

The avoidance of the ether, a subject where Newton's speculations were, to say the least, embarrassing for any apologist wishing to point the contrast with Cartesian systems, has already been mentioned in connection with Voltaire. The vogue for imponderable fluids was to come later. But already in 'sGravesande's text we can see those assumptions about the internal structure of matter, and the forces underlying chemical behavior, which were increasingly to infiltrate chemical thinking and research. We shall say more about the subject of fire when considering Boerhaave, as 'sGravesande's exposition appears to follow his teacher's closely. Here it is sufficient to note the wide connotations Newtonian natural philosophy enjoyed even at this early stage. In the first systematic textbook of the subject, 'sGravesande apparently thought it quite natural to devote three whole chapters to fire.

[33] See 'sGravesande (1720b), 13, 15, 16–23. For the Newton quotation, see note 33 of Chapter 2 above. The writer is indebted to Professor R. E. Schofield for drawing his attention to the passage in Rowning (1735), *Natural Philosophy*, II, 5–6.

The Spread of Newtonianism

4.3.2. Musschenbroek

The Dutch tradition of Newtonian exposition was also taken up by Peter van Musschenbroek.[34] Like 'sGravesande, Musschenbroek was a student at Leyden and, following 'sGravesande's return to Holland from England in 1717, the two seem to have become firm friends. Musschenbroek occupied chairs of mathematics at Duisberg (1719), Utrecht (1723), and finally Leyden (1739). His experimental bent and Newtonian propensities were well displayed in his 1729 *Physicae Experimentales, et Geometricae, de Magnete, Tuborum Capillarium Vitreorumque Speculorum Attractione, Magnitudine Terrae, Cohaerentia Corporum Firmorum Dissertationes* . . . , a work which made clear his debts, and sources of inspiration, in Newton, Hauksbee, and Taylor. We have already mentioned the French edition of Musschenbroek's 1736 Dutch textbook. His later *Elementa Physicae* appeared in English translation in 1744, while his *Introductio ad Philosophiam Naturalem* was published in Paris in translation in 1769, in time to swell the growing tide of French Newtonianism.

Limiting ourselves solely to the 1744 English *Elements,* we may note the usual discussion of infinite divisibility, and the rather "hard-line" conclusion that experiment "seems in some measure to prove, that bodies cannot be divided *in infinitum.*" Indeed Musschenbroek came out perhaps more strongly than any previous textbook writer for an atomistic interpretation of Newton, saying: "This doctrine of atoms is very ancient, and has been cultivated by Moschus, Leucippus, Democritus, Epicurus, Lucretius, Gassendus, Newton, Boerhaave, Desaguliers and others; who suppose atoms to be very small corpuscles, of which all the greater bodies are composed." Further discussion led to the conclusion that "we are ignorant whether these least solids are of the same or a different magnitude . . . Whether

---

[34] (1692–1761). See Michaud's *Biographie Universelle,* and Brunet (1926), *Les Physiciens Hollandais.* See also *JB,* XII, 618–619, for how in 1726 the Royal Society received "a small discourse in Latin from Dr Musschenbroek at Utrecht . . . Dr Musschenbroek was ordered thanks."

they are like or unlike to one another? What is their real magnitude?" However on the question of their union to form "particles of the first order" and so on upwards in the approved Newtonian manner, Musschenbroek was thoroughly orthodox.

The exposition of attractive and repulsive forces is noteworthy for its host of chemical examples, its fulsome references to Boerhaave's chemistry, and its exposition of Brook Taylor's attraction experiments. Once again the ether is conspicuous only by its absence. Of the chapter on fire we may just mention Musschenbroek's opposition to Boerhaave's ideas on whether fire has weight, and, as an example of the degree of sophistication obtained in a textbook of natural philosophy—not chemistry—by the mid-century, the one particular argument that

> another difficulty has been moved, that bodies, which have been calcined by a burning-glass, acquire a greater increase of weight in open vessels, than in those which are closed; therefore they say, that some particles have accrued from the air, which have increased the weight. But if in closed vessels the weight of the calcined bodies has increased, this could not be owing to the air, the access of which was hindered. The greater or less increment depends upon a longer or shorter, a stronger or lighter calcination, and therefore we must conclude, that fire is endued with gravity.[35]

As this one example makes abundantly clear, Newtonian textbook writers in the Dutch tradition believed chemistry and chemical problems to lie within their sphere of interest. But to see a commitment to Newtonian philosophy united to a really deep chemical insight, we must turn to the work of Boerhaave himself.

4.3.3. Boerhaave

We have already mentioned Herman Boerhaave's[36] contact with Archibald Pitcairne, Richard Mead, and other Scottish physicians. Boerhaave's long reign at the University of Leyden

[35] Musschenbroek (1744), I, 19, 21, 25, 32–36, and II, 18–19.
[36] (1668–1738). See Burton (1746), *Life and Writings of Boerhaave,* and the series of studies by G. A. Lindeboom (1959 and 1962–64).

and his commanding position in chemistry, botany, and medicine were to be of the utmost importance, not only for the spread of Newtonian doctrines on the Continent, but also for the whole development of Scottish medical education. When we recall that both William Cullen and Joseph Black gave their chemical lectures primarily to medical students, that John Dalton aspired to and William Henry actually did take the medical course at Edinburgh, and that "during [Boerhaave's] tenure [at Leyden] there were 1919 students in the faculty of medicine" (no less than 659 being from Great Britain and her colonies),[37] we can begin to understand the importance of Boerhaave to the course of chemical theory.

It may perhaps be relevant to observe here that one of the most important facts about the development of chemistry in the eighteenth century was its gradual professionalization. Instead of being an appendix to such diverse studies as medicine, metallurgy, pharmacy, and natural philosophy, chemistry was slowly establishing itself as a subject in its own right—a subject, that is, in which paid, full-time teaching and research posts were available. That this development was intimately bound up with the growing importance of chemical manufactures is at once both obvious and unexplored. The development was of course also closely linked to the growing importance of chemical medicines and the rising demand for medical education throughout western Europe. Thus while the few "professional" chemists living were almost all associated with university medical faculties at the start of the century, an increasing group may be found in a manufacturing and "popular lecturing" context as the century wears on. Even so it would be a serious anachronism to think of almost any figure before the late eighteenth century as purely and simply "a chemist."

A major concern of this present essay is to show how one approach to chemistry—that of Newtonian natural philosophy—became woven into the thinking of the emerging class of professional chemists through the period of the early industrial

---

[37] Quoted from Kerker (1955), "Boerhaave and pneumatic chemistry." For further information, see Smith (1932), *English Medical Students at Leyden.*

revolution. This is not for a moment to suppose that there were no other intellectual traditions making contributions to the formation of a coherent professional discipline. Indeed we shall argue that some of these other traditions were closer to chemical experience and had more of value to offer, both to the theoretician and the manufacturer, than did the all-pervasive and prestigious Newtonian philosophy. The immediate relevance of Boerhaave to our purpose lies in the fact that, in the period when medical teaching and practice still exercised a dominant influence on chemistry as a university subject, his writings were to integrate Newtonian natural philosophy and the traditional medical concern for chemical knowledge and chemical remedies.

Among the many works attributed to Boerhaave, some spurious, some authentic, one of the earliest and most delightful, and one which clearly displays his Newtonian sympathies, is the 1719 *Method of Studying Physick*. This book is also of the highest value for its discussion of a host of writers on every subject remotely connected with medicine, its pithy contemporary comments on their reliability and worth, and its long list of authors and works. Boerhaave's Newtonian sympathies are most evident in the section on physics, where not only is Descartes not among the listed authors, but we are told in no uncertain terms that "the prince and captain of all is Sir *Isaac Newton,* who knows as much as the rest of mankind together . . . An example of this we have in his first volume in the chapter of opticks and colours. I never saw a book where were stronger arguments drawn from experiments: It is the best pattern in the world and deserves the highest honour." After this it is a comparative anticlimax to read of chemistry that "this science in the first place discovers what are the powers of every single body . . . which can never be known but by experiments . . . Sir Isaac Newton gives us many chymical experiments of the attraction of bodies."

Boerhaave's reconciliation of infinite divisibility, atomism, porosity, and gravity is at once so masterly, so "Newtonian," and so lucid that it deserves quotation:

The Spread of Newtonianism

This cause of cohesion, a thing most unknown in the world . . . I shall distinguish . . . as *attractick virtue* with Sir Isaac Newton . . . If this cohesion or hardness in any part is such as exceeds all known power of separation, it is called an *atom* or corporeal element . . . The quantity of corporeal, impenetrable matter in every space is known by its weight, so that if we could have only one body perfectly known to us, we might explain all bodies, but there is no such thing in the world: gold seems to be solid, but it is demonstrated . . . to be porous, therefore our science can only be known in so much as it is relative . . . Hitherto we have seen that an indivisible consists of infinite parts, which being united by some power make an atom, which also has gravity, and therefore it is evident that extended, impenetrable matter, formed into certain parts which cannot be changed by any power whatever, have a gravity perfectly proportionate to the gravitating mass.[38]

This discussion was to be repeated and amplified in the later *Treatise on the Powers of Medicine*.

An interesting side light on Boerhaave's Newtonianism is to be found in a letter of his to Fontenelle in 1725. Boerhaave wrote that "after elastic force perhaps *magnetic* attraction can be held to be the other great instrument of living nature. [The study of] this force, cultivated by the authors of antiquity and restored by the most exalted Newton's acceptance of it, [now] delights our generation. Tracing this everywhere in various species of bodies, I grasp that they have it partly in common and indeed partly individually."[39] Fontenelle's reaction is not recorded!

To see Boerhaave's Newtonianism fully united to his chemical knowledge, it is necessary to consult his classic textbook, the *Elementa Chemiae*. Boerhaave's reluctance to publish a chemical text, and the resultant appearance of spurious editions based

---

[38] Boerhaave (1719), 98–101 and 19–26. The themes of the last quotation will be returned to in section 6.5. below.

[39] Lindeboom (1962–64), *Boerhaave's Correspondence*, no. 147.

Atoms and Powers

on notes of his lectures, is well known and fully documented.[40] Indeed the 1727 English edition of his *Chemistry* produced by Peter Shaw and Ephraim Chambers was, despite the claims of its title page, an unauthorized work. Boerhaave was finally driven to produce his own version of his lectures, in self-defense. The genuine *Elementa Chemiae* appeared in 1731. An authorized English translation, undertaken by Timothy Dallowe with Boerhaave's assistance, was published in 1735.

Though far more restrained than the version of Shaw and Chambers, Dallowe's authorized text still made clear to the reader the importance of Newton. Thus we have "the singular penetration of the incomparable Newton" (p. 132), "the great Newton" (p. 152), "the illustrious Newton" (p. 165), and "the great Newton, whose uncommon genius seems to have penetrated almost beyond the limits of human understanding" (p. 232). If all this was not sufficient to make Boerhaave's allegiance plain, the very arrangement of the first volume ("theory of the art") with its "experiments," "scholia," "corollaries," and "queries" could not but remind the contemporary reader of Newton.

As a good Newtonian, Boerhaave was convinced that there were in Nature "certain corpuscles" incapable of change or division. Of course these "exceeding fine elements" were far below the observational level, and as such not to be confused with the substances with which chemists dealt:

> These principles the philosophers call, elements of bodies; and into these the chemists have ofen asserted that they have reduced them; but they themselves confute their own opinions . . . And indeed, that there is nothing of this simplicity in the common chemical operations, has long ago been past dispute . . . what person living can by any art ever shew us one drop of pure simple water? . . . Nay farther, the parts into which the greatest masters pretend to have resolved compound bodies, are not themselves of a simple nature, but mutable and capable of farther division. This the water,

[40] Gibbs (1957), "Boerhaave's chemical writings."

spirits, salt, oil, and earth extracted from animal, or vegetable bodies plainly evinces, nay alcohol itself is separated in burning into different principles.[41]

In this passage, which is of the greatest importance to our understanding of many eighteenth-century confusions, we see compounded the practical difficulties and Newtonian philosophy which together were so to influence the course of chemical research. The practical uncertainty as to what were the chemical elements sprang of course from Robert Boyle's critique. But, as has been pointed out in the case of Boyle himself, a corpuscular philosophy was basically subversive of chemical theory.[42] Particularly in the form it received in Newton's hands, this philosophy denied privileged status to any set of chemicals, suggesting rather that all the materials which the chemists handled were highly complex. The belief in the complexity of all chemicals ("the internal structure to matter") is one which has already been discussed, and to which we shall return. Here it is sufficient to note Boerhaave's use of this belief.

One of the longest and most influential sections of the *Elementa Chemiae* was its masterly exposition of the crucial subject of fire. This exposition was to be of the greatest significance to the development of Newtonian natural philosophy. Early-eighteenth-century thought on heat, light, flame, and combustion demands at least a monograph of its own, and not until such a monograph has been provided shall we be able to judge the true significance of the phlogiston controversy. However the importance of Boerhaave's exposition to this present study lies rather in the use it made of attractive and repulsive forces, and in the strong resemblance between the properties of "fire" and of Newton's ether.

Boerhaave argued that "all the observations that I have had an opportunity of making seem to evince, that where there is neither any degree of attrition, nor motion from the mixture

[41] Boerhaave (1735), I, 46–47.
[42] See Kuhn (1952), "Boyle and structural chemistry."

of various bodies together, there, fire is most equally distributed through every part of space." This ability of fire to distribute itself through space, even when the space was occupied by bodies, arose from the porous nature of substances. Thus "fire is not able to insinuate itself, into what we call the ultimate impenetrable elements of body, but that it is repelled thence whenever it exerts itself upon them . . . fire itself does in reality never reside in the proper substance of bodies, but only in the vacuities." [43]

The resemblances between the properties of fire, as here developed, and the properties of the Newtonian ether are immediately apparent. This is surprising when we recall how little stress was placed on, or use made of, the ether before the 1740's. It would be intriguing to know whether Boerhaave was aware of the early speculations of Newton, as contained in the *Letter to Boyle* for instance. A more likely explanation, in view of the catholicity of his thought, is that Boerhaave was utilizing those parts of his Cartesian heritage which he felt blended most satisfactorily with his newer commitment to Newton, his passion for natural history, and his great experimental ability. Certainly Boerhaave's doctrine of fire, whatever the influences behind its development, was to find a sympathetic audience among the ether-conscious experimenters of the fifties and the sixties. Indeed it has been argued that Benjamin Franklin's important "one fluid" theory of electricity owed much to it.

There is not time further to expound Boerhaave's careful and subtle arguments and experiments on fire. His belief in the "conservation of fire," his treatment of the relationship among fire, flame, and inflammability, his discussion of the role of air in combustion, and his carefully argued rejection of Boyle's belief that fire possesses weight—all these passages deserve reading for their clarity and careful argument, and pondering for their influence on eighteenth-century natural philosophy.

It is difficult to overestimate the importance of Boerhaave's *Chemistry* for the subsequent development of chemical thought. In its Latin version and in a host of translations into a variety

---

[43] Boerhaave (1735), I, 113–115.

of languages, it was—by its sheer size and scope as well as its lucidity and authority—to exercise a dominant influence for at least half a century. As Richard Mead so aptly put it: "Your work on chemistry . . . by Jove, is learned and must have been an arduous task . . . it is worth its weight in gold." At a far later date, Dalton's statement that "with regard to chymistry, I sometime since perused Boerhave's treatise which I suppose is a capital one; making allowance for the time since it was written" was at least as much a commentary on the enduring value of Boerhaave's work, as on Dalton's own lack of chemical instruction.[44]

The writings of 'sGravesande, Musschenbroek, and Boerhaave served not only to establish Newtonian doctrines in Holland but also to aid their diffusion through Europe, and to demonstrate their relevance to chemical discussion and research. The wide popularity which their textbooks enjoyed was not least apparent in Britain, where translations served to complement the flourishing home production of Newtonian texts. And the close relationships that prevailed between Dutch natural philosophers and the Royal Society were still further strengthened by Desaguliers's 1731 and 1732 lecture courses in Holland.[45]

4.4. Newtonianism Established in Chemistry

The final part of this discussion will deal not with standard Newtonian texts, but rather with the few second-generation

---

[44] For Mead's comment, see Lindeboom (1962–64), *Boerhaave's Correspondence*, no. 125. This letter would seem to testify to the veracity of Mead's biographer, who referred to "a constant correspondence" existing "between Dr. Mead and Boerhaave, who had been fellow students at Leyden. They communicated to each other their observations and projects, and mutually gave and received presents": Mead (1755), *Authentic Memoirs,* 27–28. Such a correspondence between two so influential men, if rediscovered, would provide fascinating insights into the mechanism of Anglo-Dutch Newtonian cooperation. For Dalton's comment, see Brockbank (1929), *Dalton, Physiologist,* 7.

[45] Desaguliers's visits to Holland are mentioned in the biographical accounts, e.g., Marchand (1758–59). Documentary evidence for 1731 and 1732 is now available in a 29 October 1733 letter from Desaguliers to Cromwell Mortimer, secretary of the Royal Society: see Royal Society Ms. D 2 73. Further information on the Desaguliers–'sGravesande relationship is much to be desired.

British Newtonian works of more definitely chemical orientation. Perhaps the most surprising thing about the chemical work of the early British Newtonians is how little it was exploited by later British workers. Hauksbee's electrical experiments were, through their extension and systematization by Gray, du Fay, and Desaguliers, to lead to the vigorous speculation and research of the 1740's, 1750's and 1760's.[46] This research was pursued as actively in Britain as anywhere. But the chemical work of the Keills, Freind, and Taylor was to be exploited best on the Continent, and especially in France. The reasons for this will be more fully explored in Chapter 7. Here our concern is with British developments of Newtonian chemistry up to the 1740's, developments most clearly associated with Stephen Hales, Peter Shaw, and Henry Pemberton.

Stephen Hales[47] went up to Cambridge as a nineteen-year-old undergraduate in 1696, the year Newton was appointed to the Mint. As a student and latterly as a fellow of his college (Bene't, now Corpus Christi), he remained in Cambridge for almost thirteen years. His friendship with William Stukeley, and their joint pursuit of a wide range of scientific topics, is well known. It is more difficult to provide evidence of Hales's awareness of the early Newtonians, but as a Cambridge natural philosopher he cannot have been unaware of Roger Cotes or William Whiston. And when Hales finally appeared at the Royal Society in 1717, he was obviously no stranger, nor was his work.

Hales must have known and argued with John Waller and John Mickleburgh, two other members of his college who in turn (1708, and 1717) succeeded to the chair of chemistry. And the notes of one of Mickleburgh's courses, tentatively dated to 1726, show the strong hold of Newtonianism by that time: "The first who applied Sr Isaac Newton's philosophy to chymistry was Dr. Freind and how happy and successful he hath been in this application every one who hath or will but read his chymical

---

[46] On which, see Cohen (1956), *Franklin and Newton*.
[47] (1677–1761). A full-length biography is available: see Clark-Kennedy (1929).

The Spread of Newtonianism

lectures will be able to judge." The long-term importance of this Cambridge chemical Newtonianism is placed in a somewhat different light when we remember the torpor into which the English universities were sinking at this time. Mickleburgh apparently gave only five courses of lectures in all his thirty-eight years as professor, and there were no chemistry lectures at all in the seventeen years following 1741.[48]

The greater part of Hales's own work must have followed his removal from Cambridge to become minister of Teddington, near London. The earliest record of his attendance at the Royal Society is for 14 March 1717. He was elected F.R.S. a year later. By that time the vigorous early Newtonian activity connected with the Keills, Freind, Hauksbee, Taylor, etc. had largely passed away, and even the period of most intense French curiosity was passing. Nonetheless when, with Newton in the chair, Hales on 15 March 1718 read his first paper about "the effect of the sun's warmth in raising the sap in trees," he "was desired to prosecute these experiments, and had thanks for communicating this first essay." The end result of this encouragement, and of the Newtonian orientation of the Royal Society, was to be his 1727 classic, *Vegetable Staticks*. Whole sections of this work were read before the society, and when the published version finally appeared, it bore the proud legend "Feb. 16, 1727. Imprimatur Isaac Newton, Pr. Reg. Soc."—one of the last works to do so, for Newton would be dead in just over a month.

Hales's indebtedness to the early British Newtonians was spelled out on almost every page of his book. The preface told how

> it appears by many chymio-statical experiments, that there is diffused thro' all natural, mutually attracting bodies, a large proportion of particles, which, as the first great author of this important discovery, Sir Isaac Newton, observes, are capable

---

[48] See Coleby (1952b), "John Mickleburgh." Quotation from p. 168, citing Caius College Cambridge, Ms. 619/342 red. Five whole pages of the manuscript (two and a half lectures) are devoted to Freind, Newton in Query 31, and the chemical importance of attractive and repulsive forces.

of being thrown off from dense bodies . . . into a . . . permanently repelling state . . . It is by these properties of the particles of matter that he solves the principal phaenomena of Nature. And Dr. Freind has from the same principles given a very ingenious rationale of the chief operations in chymistry.[49]

The tone thus set was maintained throughout the work, with references to James' Keill, Hauksbee, Shaw's newly published version of Boerhaave, and of course to Newton himself (the *De Natura Acidorum,* and Queries nos. 7, 9, 10, 18, 21, 29, 30, and 31 in *Opticks*.)

The importance of Hales's work on "air," and its influence on later Continental chemists, is now generally recognized. Hales's clear demonstration of the part "air" plays in chemical processes was not only the starting point of much subsequent fruitful research, but also had a great effect during the next half century on thought about the chemical elements. The immediate impact of his work may be seen in Shaw's addition of air to the list of chemical principles, and in the comments in the authorized version of Boerhaave's chemistry, while the wide influence of the *Vegetable Staticks* is revealed in the variety of languages into which it was translated.[50] However the primary importance of Hales's book to this study lies neither in its prac-

---

[49] Hales (1727), v. The Royal Society's *Journal Book* for 14 January 1725 records how "the reverend Mr Hale communicated a treatise concerning the power of vegetation consisting of six heads of experiments in six chapters. Part of the first was read and the rest was ordered for the next meeting. . . [the society desired] him to continue in the prosecution of a design which seems to promise so fairly for making an advancement in natural knowledge." On 21 January "Mr Hales was desired . . . to permit the [work] to be made public." On 18 February "the Society repeated their thanks to Mr Hale for communicating his curious experiments," while on 4 March they again "desired him to print it." Further expressions of enthusiasm for such a brilliant and Newtonian work occurred in 1727, when Hales returned with his sixth chapter on air. The reading of this occupied the society for the whole of February and March, until interrupted (23 March) by news of Newton's death: see *JB*, XII, 524, 526, 535, 537–538, etc., and XIII, 44, 45, 48, 51–62, etc.

[50] Including Dutch in 1734, French in 1735, German in 1748, and Italian in 1756. See Guerlac (1951), "Continental Reputation of Hales." See also Shaw (1734b), *Chemical Lectures,* *147, and Boerhaave (1735), *Chemistry,* I, 314.

tical demonstration of the chemical role of the atmosphere, nor in the undoubted brilliance of its investigations into plant physiology. (These latter were of course the real reason for the book, just as animal physiology was the subject of his later *Haemastaticks*.) Its importance lies rather in three related facts: that it drew attention to the Newtonian tradition, that it demonstrated how richly fruitful the exploitation of Newton's Queries could be, and that it showed how repulsive forces were of the greatest utility for Newtonian natural philosophy in general, and chemistry in particular.

As regards the first of these points, it is sufficient to observe the many admiring readers Hales enjoyed, up to and including Lavoisier.[51] These readers could not fail to notice the value he placed on the early British Newtonian group—and thus, perhaps, be persuaded to read Freind and his associates for themselves. As to the second, almost half of Hales's book (162 pages) was devoted to his "specimen of an attempt to analyze the air." Yet this inquiry took its rise from a few brief statements about "true permanent air" in but three of Newton's queries. Small wonder that Hales himself should say: "I hope the publication of this specimen of what I have hitherto done, will put others upon the same pursuits, there being in so large a field, and among such an innumerable variety of subjects, abundant room for many heads and hands to be employed in the work: For the wonderful and secret operations of Nature are so involved and intricate." Small wonder either that Desaguliers should refer in 1734 to "the Rev. Mr. Stephen Hale's excellent book *Vegetable Staticks*, which, by putting several of Sir Isaac's Queries out of all doubt, shew how well they were founded."[52] Nor is it surprising that Buffon, convinced of the value of Newton's work, should begin the education of his countrymen by translating *Vegetable Staticks*.

With regard to the third point, we have seen how Newtonian

---

[51] His early reading of Hales is now established by Gough (1968), "Lavoisier—new evidence."

[52] Hales (1727), vii; Desaguliers (1734–44), *Experimental Philosophy*, I, sig. a 4^v.

texts, such as 'sGravesande's *Physices Elementa,* referred to but did not stress the role in Nature of repulsive forces. This is understandable as, though the Queries in *Opticks* and his ethereal speculations did include repulsive forces, the stress of Newton's work was on attraction. In the *Principia,* the mutual repulsion of air particles was presented only as a possibility, and the whole stress of the work was on the mathematics of gravitational attraction. Similarly the *Opticks* was primarily concerned with the mutual attraction of light and gross mattter. But with Hales, the existence of a repulsive force became an essential part of the economy of Nature. As he pointed out:

> Air abounds in animal, vegetable and mineral substances; in all which it bears a considerable part: if all the parts of matter were only endued with a strongly attracting power, whole Nature would then immediately become one unactive cohering lump; wherefore it was absolutely necessary, in order to the actuating and enlivening this vast mass of attracting matter, that there should be everywhere intermixed with it a due proportion of strongly repelling elastick particles . . . that thereby this beautiful frame of things might be maintained, in a continual round of the production and dissolution of animal and vegetable bodies.[53]

With this clear demonstration of the importance of the repulsive force to chemistry, we have one more step toward the formation of that philosophy which was to underlie so much chemical thought and experiment in the latter part of the century.

Though Hales lived to a ripe old age, none of his later work concerns us—and indeed none reached the outstanding level of his 1727 book. The other British Newtonian and chemist of note in this period was Peter Shaw,[54] and he was more remarkable as a translator, epitomizer, and abridger than as an original

---

[53] Hales (1727), 313–314.
[54] (1694–1763). See Gibbs (1951b), "Peter Shaw."

thinker. This does not mean that he can safely be neglected. He is important because the eighteenth-century British student of Boyle, Bacon, Boerhaave, or Stahl was more than likely to use a version with preface, introduction, and, if need be, translation by Shaw. For instance the reader of his 1730 translation of Stahl, the *Philosophical Principles of Universal Chemistry* (still the only complete work of Stahl's available in English), will search in vain for any mention of phlogiston!

Shaw, about whose early life little is known, was firmly inside the medicochemical tradition, from which so much of the early British Newtonian work had sprung. As early as 1723, in his *Treatise of Uncurable Diseases,* he was maintaining that "there was room for the application of the Newtonian method in medicine," [55] while his 1734 *Enquiry into . . . Scarborough Spawwaters* was dedicated to Richard Mead. The preface to his head "chymistry" in his 1725 three-volume methodized abridgment of Boyle's works was also full of Newtonian enthusiasms. This preface quoted at length, and approvingly, Boerhaave's 1718 inaugural address on the study of chemistry, itself a work of marked Newtonian sympathy.

Shaw's 1731 *Scheme for a Course of Philosophical Chemistry* contains references to Freind's *Praelectiones Chemicae* and the *De Natura Acidorum,* as well as such subject headings as "A view of the different relations vulgarly call'd sympathies and antipathies, or attractions and repulses, observ'd betwixt different chemical bodies; with the uses of this doctrine in philosophy and chemistry. See Boyle, Hook, Homberg, Newton, Stahl, and the memoir of Geoffroy in the works of the Royal Academy for the year 1718." [56] As the foregoing illustration makes plain, Shaw was not afraid to "integrate" the works of such diverse authors as Stahl, Homberg, and Newton. The nature of such

---

[55] Gibbs (1951b), 213, in comment on Shaw (1723), which the present writer has not seen. See also Shaw (1725), *Boyle's Works,* III, cclvi, cclix–cclx etc.: e.g., "Let any man compare the *Praelectiones Chymicae* of Dr Freind, with the common *Tyrocinia* of chymical writers . . . and he will be sensible of . . . [the] difference."

[56] Shaw (1731), 41.

integrations, which were to become increasingly important as the century progressed, will be considered briefly in Chapter 6.

One final claimant to the status of British Newtonian chemist of the 1720's and 1730's must be included here. Though he is not usually thought of in that connection, Henry Pemberton [57] gave a series of public lectures on chemistry. Pemberton's career illustrates once again the importance of Boerhaave's Leyden teaching and Newton's own immediate circle in promoting and directing the popular acceptance of Newtonianism. Having studied under Boerhaave, Pemberton was initially brought to Newton's notice by John Keill, and subsequently by Richard Mead. This latter introduction led to his being charged with the production of the third (1726) edition of the *Principia*.[58] Shortly after its appearance, and even more shortly after Newton's death, Pemberton was to publish the first really vulgarized introduction to Newton's work, in the form of his eminently successful 1728 *View of Sir Isaac Newton's Philosophy*.

Though we cannot tell how valid was his claim that "Newton approved of the . . . treatise, a great part of which we read together," fashionable society was obviously willing to accept his qualifications as interpreter of "the great Sir Isaac Newton." [59] Not content with writing the *View*, Pemberton also gave popular lectures on chemistry, like Shaw. These lectures were only published posthumously in 1771. The original manuscript now being lost, it is impossible to tell how faithful the printed version was to Pemberton's actual course. Certainly the published lectures are tedious and uninspired. What is of significance here is that in the 1730's Pemberton should be seeing the whole of chemistry as explicable in terms of Newtonian ideas. However the developing British tradition of popular Newtonian lectures must be left for later consideration.

[57] (1694–1771). See *DNB*.
[58] See Pemberton (1771), *Course of Chemistry,* vi, ix–x, and xiii; also Cohen (1963), "Pemberton's translation of the *Principia*," *passim*.
[59] Pemberton (1728), sig. a 2ᵛ. The acceptance that Pemberton enjoyed is apparent from the list of over 2000 subscribers in the front of the 1728 *View*.

The Spread of Newtonianism

4.5. Some General Remarks

In this and the previous chapter, we have outlined the diffusion of Newtonian ideas in France, Holland, and Britain, and their—rapid or slow—acceptance into the "tacit assumptions" of the period. This acceptance was bound up first with the political and social pressures working to produce a definite Newtonian "circle," centered on the early eighteenth-century Royal Society, and later with the development of a strong tradition of Newtonian textbooks of natural philosophy. These textbooks served to expound, explain—and if need be to extend—the work of the master. From what has already been said it should be apparent that they all showed a definite, if limited, awareness that chemistry lay within the province of Newtonian natural philosophy.

Within the broad tradition of Newtonian approaches to chemistry, we may perhaps distinguish three strands. One is a formal, mathematical concern with the science—exemplified in such diverse figures as (John) Keill, Freind, and Maupertuis (and, later, Clairaut, Buffon, Cavendish, Richter, Bergman, and Laplace). The members of this tradition, by no means necessarily "positivist" as opposed to "speculative," saw Newton's great achievement as the *mathematization* of the heavens in terms of a $1/r^2$ law, and sought a similar mathematization of earthly phenomena. Their conviction that there was a chemical force law, and that chemistry could therefore be mathematized like astronomy, was to provide much of the driving force behind late-eighteenth-century affinity studies. The second strand is a less formal and less mathematical one, essentially deriving from a medicochemical concern with chemical phenomena, and the wish to integrate this experimental interest and the approved (i.e., Newtonian) natural philosophy. This strand is exemplified by such medically qualified figures as (James) Keill, Mead, Geoffroy, Boerhaave, and Shaw (and, later, Cullen, Black, Fourcroy, and Berthollet). With them the stress lies rather on the experimental investigation of chemical forces than on their abstract mathematization. The third, and in many ways most interesting strand, is that peculiarly British tradition of popular lecturers, from Harris through Desaguliers and Martin to Garnett, Higgins, and Walker. This tradition will be considered briefly in Chapter 8.

Though we may distinguish three strands in chemical Newtonianism, members of all were agreed in their acceptance of a common theory of matter. The theory had two essential components—belief in the inertial homogeneity of all matter and its possession of an "internal structure," and acceptance of attractive and repulsive forces as proper categories of explanation.

As we have tried to show, belief in the internal structure of matter was common to all mainstream Newtonians. Its basis lay partly in the demand of Newton's optical theory for highly porous matter, partly in theological concerns, partly in inherited "corpuscular" assumptions, and partly in considerations of simplicity assumed but not made explicit in the *Principia*. The relevance to chemistry of attractive forces was clear from Newton's writings, but the widespread realization of the equal importance of repulsive forces may be dated from Hales's book. And the peculiarly eighteenth-century "business of experimental philosophy" was to be the discovery and quantification of all the forces that existed, not least those involved in chemical reaction.

A third and more ambiguous Newtonian category, the ether, though often referred to or hinted at, did not feature prominently before the 1740's. Its unfortunate resemblance to the Cartesian dense fluid ether, which Newtonians wished to discredit, was one reason. And Newton's own hesitations were no doubt another. However both Boerhaave's "fire" and Hales's "air" bore remarkable similarities to Newton's own subtle ether. Their obvious explanatory utility must take its place along side the growth of electrical studies in the late 1730's, as a reason for the increasingly widespread interest in the Newton-

The Spread of Newtonianism

ian ether evident in the 1740's. Foremost among the new exploiters of the ether were the British representatives of the emerging group of Newtonian systematizers we must now consider.

5
Speculative Systems

> I really think that there can be found in this *Theoria* the general theory for all chemical operations.
> R. J. Boscovich in 1763

5.1. Introduction

To the eighteenth century, the fascination of Newton's work did not lie only in the key it seemed to offer to the unlocking of Nature's secrets. The fascination lay also in the unfinished, enigmatic, puzzling character of the work itself. "A key to unlock the innermost sanctuaries of Nature": what more could one wish for? In the 1720's and 1730's, when the primary task was winning allegiance to Newton and writing those first-generation textbooks which would instruct believers in the new Newtonian categories, the answer was "very little." But in the 1740's and 1750's, with the pressure of strange new experimental results mounting, and inconsistencies between different parts of the master's written word becoming more apparent, the need was rather for a key to Newton—a guide to show the true nature of the Newtonian frame, and hence the most fruitful road for further studies.

The task of systematizing Newton was to attract some of the mid-century's finest, and many of its lesser, intellects. Here we can only sample the different styles of those speculative Newtonian philosophies which endeavored to create coherent and logical deductive accounts of the whole of Nature. These systems were based on the common foundations of the inertial homogeneity of matter, the existence of attractive and repulsive forces, and the possibility and desirability of a mathematico-physico explanation of all natural phenomena. Such schemes were to exercise an enormous influence on scientific thought later in the century. Though directly eliciting no new facts, and leading up to no crucial experiments, speculative force theories were to condition the whole atmosphere of chemical research.

It was among the second and third generation Newtonians of mid-century Britain that the creation of speculative systems reached its peak. With so many startling electrical discoveries being made in the Anglo-Saxon world, the need to accommodate new knowledge within the Newtonian frame was especially acute. At this juncture Newton's previously neglected ether speculations were to prove of particular utility. It is thus no mere accident that Bryan Robinson's earlier and more muted thoughts burst into full bloom in 1743 in the *Dissertation on the Aether of Sir Isaac Newton,* nor that the enterprising Thomas Birch was soon after publishing to the world Newton's own highly suggestive *Letter to Boyle.*

Activity, and especially the invocation of imponderable fluids, was greatest in Britain. However, the same wish to systematize Newton may be seen among Continental authors, though at slightly later dates. The most influential scheme of all was that produced in 1758 by the widely traveled Croatian Jesuit R. J. Boscovich, while the rather different ideas of the Comte de Buffon were to have a marked influence on French chemists. Interestingly, neither of these latter systematists had any role for the ether—perhaps a reflection of the Continental's still-felt need to reject Descartes completely if he wished to accept Newton. After all, Cartesianism, though "almost extinct," was not entirely so: the period of Franklin's thoroughly Newtonian imponderable fluid theory of electricity was also that in which, in France, the Abbé Nollet was vigorously advocating neo-Cartesian explanations of the same phenomena.[1]

---

[1] See Cohen (1956), *Franklin and Newton, passim.*

5.2. Robert Green and Expansive and Contractive Forces

An exact contemporary of Stephen Hales at Cambridge, Robert Green[2] graduated B.A. in 1699 and M.A. in 1703. His remaining twenty-seven years were spent quietly as a fellow of his college (Clare), and he is known to us chiefly through his three books. Of these, one dealt with theology and two with natural philosophy—a distribution characteristic of the early British Newtonian circle. But, while obviously aware of their doings, Green was far from being a member of that circle. In fact his two philosophical works had as a primary aim the overthrowing of the Newtonian philosophy. A strict Anglican, Green was even more concerned about atheism than Newton was. He therefore wished to banish materialistic corpuscular philosophy in all its forms—not only Newton's newer version, but also that of gentlemen "of a much lower class, Mr. Hobs Mr. Lock and Spinoza."[3]

Green was widely read in the standard authors of the period, as the continual marginal references in his books make clear. He had a clear grasp of writers as varied as Lemery and Freind (indeed his writings are of great interest for their information on works available and studied in Cambridge in the early eighteenth century).[4] Unfortunately Green was by no means so clear and lucid in setting forth his own views as in attacking those of others, which may account for the poor hearing given his ideas, both in his lifetime and later. Typical of the reception accorded him by the early Newtonians was Roger Cotes's comment to a friend:

> I think that [Newton's] philosophy needs no defence . . . one Mr. Green, a fellow of Clare Hall in our university, seems

[2] (1678?–1730). See *DNB*.

[3] Green (1712a), *Natural Philosophy*, sig. a 2ᵛ. And see Green (1711), *Truth of Christian Religion*.

[4] Further valuable material on this subject is available in Wordsworth (1877), *Scholae Academicae*.

to have nearly the same design with those German and French objectors, whom you mention. His book is now in our press, and is almost finished. I am told he will add an appendix in which he undertakes also to square the circle. Ex pede Herculem; I need not recommend his performance any further.[5]

Despite such hostile appraisals, Green's writings deserve our attention, if only for the way they attacked Newtonian belief in homogeneous matter and a vacuum, seeking rather to explain the observable world in terms of heterogeneous matter and "expansive and contractive forces." His *Principles of Natural Philosophy*, published in 1712, had the subtitle "In which is shewn the insufficiency of the present systems, to give us any just account of that science: and the necessity there is of some new principles, in order to furnish us with a true and real knowledge of Nature." The contents alone ran to forty pages, as Green included extensive summaries of each chapter, and the verbiage of his writing is such that it is only too easy to lose track of his positive arguments.

Green began by considering whether external nature really existed, reviewing the arguments of Descartes and Locke at great length, and with some subtlety. The question having been answered affirmatively, he went on to show his opposition to both Cartesian and Newtonian theories of matter by saying, "we cannot allow extension an existence, separate from every thing else, nor yet that it is the essence of body; and therefore, tho' we deny the vacuum of the moderns, we do not affirm the plenum of Descartes."

The arguments for a vacuum used in the first edition of the *Principia* came in for close attention. Green was obviously well aware of the changing nature of Newtonian thought—no doubt through his contacts with such Cambridge contemporaries as

---

[5] Rigaud (1841), *Correspondence of Scientific Men,* I, 261. For a further attack, see Green (1712b), *Philosophical Fanaticism.* That Green also earned the ire of German philosophers is apparent from *Acta Eruditorum* (1729), 241–245.

Atoms and Powers

Roger Cotes and Stephen Hales. As he put it himself in 1712: "I am convinced by experience, and from a conversation I have had with some excellent men in philosophy and mathematicks, that they now take several propositions in Sir Is. Newton himself, as invincible proofs of the certainty of a vacuum, which were not design'd by that incomparable geometer as such." And in his 1727 *Philosophy of Expansive and Contractive Forces,* Green could evidence the changes in Proposition 6, corollary 3 of Book 3 of the *Principia* (concerning the possibility of a vacuum), and the new ether Queries of the 1717 *Opticks,* as clear proof of Newton's own hesitancies and gradual movement toward the "Greenian" position! [6] If this latter claim seems folly, we may at least note that Green, unlike many later British writers, was acutely aware that Newton's ideas had changed over time.

In the *Principles,* Newton's argument for a vacuum, from the proportionality of gravitational weight and inertial mass, and observed variations in specific gravity, was stated at length, only to lead to Green's conclusion that "there are two particulars therefore to be ascertain'd in this argument, first, that all matter gravitates in proportion to its quantity, second, that all matter which gravitates in proportion to its quantity, is the same, or similar and homogeneous." In this Green was of course correct. Newton's own treatment of weight and mass had used the smuggled assumption that truly *solid* matter was of uniform inertial density, and thus that bulk was a direct measure of inertial quantity. And Green was legitimately able to reject this assumption, while also criticizing the associated notions that bulk and extension were primary properties of matter.

To be aware of problems in Newton's formulations was one thing. To escape from them into meaningful new approaches was a different problem, and quite beyond Green. The direc-

---

[6] Green (1712a), 51 and 66; (1727), 1–2. Green had presented a copy of his *Natural Philosophy* to Newton in 1712, and thus was doubly sure of his own Christian influence on so pious if misguided a man! Newton's library certainly included one, and possibly two, copies of Green's *Natural Philosophy:* see Villamil (1931), *Newton,* 78 and 106.

Speculative Systems

tion in which he sought to move may be glimpsed in the *Principles*, in such statements as the following:

> In a globe of gold, and an equal one of water, the reason wherefore the quantity of matter in both is estimated from the weight of 'em, and consequently the one to the other, as 19 to 1, is, because the matter is suppos'd similar and homogeneous in each . . . if, instead of unequal parts of the same matter, there was an equal portion of different matters in the globe of gold and that of water, their weights, which would be different where the accelerating force was the same, would not be proportional to their quantities of matter which are equal.[7]

Here, as elsewhere, Green abandoned the orthodox Newtonian view of the inertial homogeneity of matter in favor of the idea that equal bulks of solid matter might exhibit different inertial (and therefore gravitational) masses in the same situation (see Fig. I). The later sections of the *Principles* did not develop this position, but instead contended themselves with long and somewhat tedious criticisms of standard Cartesian and Newtonian explanations of such natural phenomena as reflection and refraction.

To discover what in fact Green's philosophy of matter was, we have to turn to his 1727 work, *The Principles of the Philosophy of the Expansive and Contractive Forces*. This enormous book is, once again, a disappointingly jumbled and confused work, though obviously the fruit of wide reading and long reflection. To understand its position, we cannot do better than start at the end. Here Green explained, with a clarity of philosophic understanding rare among his contemporaries, that

> there are only four possible hypotheses, which philosophers can espouse, or maintain.
>
> 1st. That of similar matter and a plenum, which is the *Cartesian* or the *French* philosophy . . .

[7] Green (1712a), 103 and 106.

2d. That of similar matter and a vacuum, which is that of *Galileus,* Sir *Isaac Newton,* and our modern philosophers . . . and is the old system of *Epicurus* compleat . . .

3d. That of a dissimilar matter and a vacuum, which has hitherto had no patrons, nor perhaps is likely to have any; because if matter is acknowledged to be dissimilar in it's own nature, there will be no occasion for vacuities, which are introduced by philosophers to difference one portion of similar matter from another.

4d. That of a dissimilar matter and a plenum, which is the system I have chosen to maintain and defend, and which, as it is properly my own, and has hitherto had no advocates for it, so I may justly call it a philosophy, which is entirely *English,* and I hope will in time appear to be the most *rational.*[8]

The first of these hypotheses did of course imply that gravity was a secondary quality, as quantity of matter (measured by bulk) was not proportional to weight. The second position, which Newton embraced, implied either that gravity was itself a primary property, or at least that it was always "associated with" matter. One or another variant of this second position commanded increasingly wide acceptance through the eighteenth century. Paradoxically, the position also became increasingly associated with the contradictory belief that there existed a "different" kind of matter, the ether, which was not subject to gravitational attraction at all. In all its variants this "Newtonian" approach, with its stress on the porosity of and complex internal structure to matter, obviously favored a belief in transmutation and discounted the uniqueness of chemical species. The existence of a "different" kind of matter, not subject to gravitational attraction, also militated against the idea that weight studies *alone* were an adequate means of investigating chemical reaction.

Dalton's adoption of the third hypothesis, which even in 1800 had "hitherto had no patrons," naturally caused disquiet among

[8] Green (1727), 934–935.

philosophically minded chemists. Not only did such a move ignore Occam's razor and a powerful, century-old Newtonian tradition, but the advantages of equating the different known chemical elements with different fundamental types of matter were by no means obvious at a time when the list of elements was undergoing drastic revision. However for the moment it is Green's development of the fourth hypothesis that must concern us. Though he obtained no converts and left no disciples, his ideas bear certain resemblances both to those developed by the British Hutchinsonians, and also to those which, through the work of Boscovich, were to gain a wide acceptance among orthodox Newtonians later in the century. (And just as with Boscovich and Newton, so with Green, the desire to reconcile natural philosophy and Christian theology led to a stress on forces at the expense of a more orthodox corpuscularian view of matter.) [9]

Green's belief in dissimilar matter and a plenum was far from simple, as force not "matter" was the basis of his philosophy. Thus "action or force in general is the essence or substratum of matter . . . it would be impossible for us to have any sensations from matter, unless by some kind of action or other impress'd upon our minds from it." And again, "if any one shall say, that mere action is inconceivable without some solid substance to support such action, and in which it should inhere; that only proceeds from our being used to consider things in the corpuscular way, which has continually inculcated to us it's atoms or particles, and the moving of them." [10] Accordingly it was on the basis of force that Green proposed to explain Nature.

Matter being heterogeneous, and force being the basis of matter, Green naturally felt free to invoke a variety of forces.

---

[9] That Green belongs in the same tradition as the Hutchinsonians is apparent in his strict Anglicanism and his denial of inertial homogeneity in matter. The Hutchinsonians will be all too briefly mentioned in sections 5.3.2. and 8.3. below: see especially note 23 to Chapter 8. Green's conservative stress on forces, like Boscovich's, was to gather very different political overtones later in the century.

[10] Green (1727), 286 and 409.

Atoms and Powers

In fact he wished to explain all observed phenomena in terms of two types of force, expansive and contractive, each of which might apparently be present to any degree in any body. For instance he said that "all bodies solid and fluid are compounded of a centripetal, a gravitating, or contractive force . . . [which] is in the same point of space, and the same instant of time, various and changeable in different sorts and portions of matter; as also of a centrifugal, elastick, or expansive . . . These forces therefore . . . are the real substratums of all bodies." What Green did not make clear was the center to or from which these forces acted and with which they were associated. Having denied the reality of "matter in itself" apart from force, he was left with the problem of explaining how his forces might be localized in space, and thus "explain" the material objects of common life. This was a problem of which he appeared blissfully unaware. Sometimes he assumed that his forces did act to or from a center while at other times they appear to be delocalized, operating and interacting through all space and giving rise, according to their proportions in any given point of space, to the appearance of various types of matter—the heterogeneous matter which filled the plenum of the universe. But of systematic, logical development of these ideas—development such as Boscovich was to undertake—there is not a trace.

The way in which Green applied his ideas, entirely nonpredictively, may be seen in a typically ungrammatical passage from his chapter on chemistry:

> Spirit of sal armoniack precipitates gold, which aqua regalis, or silver, &c. which spirit of nitre or aqua fortis dissolves; because spirit of sal armoniack has a greater expansive force and a less contractive than aqua regalis or spirit of nitre, and therefore weakens and enervates the contractive of the dissolvents, which was necessary to the making an impression upon those metals, by which their moments become too light and spirituous to support their dissolutions . . .

Speculative Systems

or again, treating gold, he said:

> It's solidity, by which we cannot penetrate into it, is owing to it's particular degree of contraction, by which the parts of it tend to one another, and to the center, as also to it's expansive force, by which it repels any force, which endeavours to enter into the place, which it possesses ... It's ductile and fusible nature is derived from its expansive, being assisted by the expansive force of percussion, or that of fire, by which it in some measure surmounts it's force of contraction.[11]

These passages, representative of great sections of the *Expansive Forces,* dealing with such things as astronomy, optics, chemistry, and anatomy, show clearly why Green's writings commanded little respect among his contemporaries. His often cogent criticisms of received doctrines, and his original ideas on explanation by means of expansive and contractive forces, were lost in a great sea of vague and confused speculation which yet purported to be an adequate and comprehensible system of Nature.

Even so it would be a mistake to dismiss Green out of hand, as most of his contemporaries appear to have done. Newton's use of the concept of force radically transformed previous mechanistic ideas about the nature and properties of matter, and posed puzzling questions for upholders of any "corpuscular philosophy." His speculations on the ether had served only to demonstrate the difficulties, both logical and practical, in providing a mechanism for force. More than that, the newer approach to epistemology—exemplified in Locke's writings and Berkeley's idealism—had created serious dissatisfaction with the too-simple assumption that men could know what were the "real" properties of matter, and especially that those "real" properties were the shape, extension, etc. of the traditional mechanical philosophies.

Green was among the first to realize that force could provide

[11] Quotations from Green (1727), 289, 317, and 289.

a more basic concept than solidity or extension when describing matter, and to seek to develop this idea. Unfortunately his imagination far outreached his power of organization and exploration, giving his work an irritating confusion. That his wish to reform the theory of matter was motivated by zeal for the defense of religious truth is wholly characteristic of eighteenth-century British natural philosophy. In his wish to replace Newton's work completely, Green represents a minor though continuing stream of orthodox Anglican thought, rather than the Socinian tradition adhered to by most British expounders, expanders, and modifiers of Newton.

Turning from the question of theological motivation to that of scientific content, we may note how, in a way rare in the eighteenth century, Green rejected prevailing belief in the inertial homogeneity of matter. Though, like Dalton at a later date, he could argue that his own ideas agreed with Newton's real beliefs, Green's writing was at once too revolutionary and too confused to affect the natural philosophers of his time. To see views in some ways similar to those he advocated meeting with response, we have to turn to the works of later deliberate systematists. Paradoxically, it was to be under the influence of one such writer, R. J. Boscovich—himself a theological conservative like Green—that half a century after the publication of the *Philosophy of Expansive and Contractive Forces* the radical theologian Joseph Priestley was to hail as boldly original the idea that matter was essentially force.

5.3. The Ether, Electricity, Magnetism, and Heat
The similarities between Boerhaave's "fire," Hales's "air," and Newton's repulsive-force ether have already been mentioned.[12] The wide diffusion of the *Elementa Chemiae,* and the obvious fruitfulness of Hales's work, helped to assure the gradual European admission of imponderable-fluid theories based on Newtonian categories of explanation. In the particular case of the

[12] See sections 4.3.3. and 4.4. above.

Anglo-Saxon world of the 1740's, there was also a much more immediate and powerful cause for the great vogue enjoyed by ether mechanisms. This cause was the literally spectacular progress being made in the experimental study of electricity. Reflecting the popular mood, Benjamin Martin was to say in 1747 that though Newton had written about an ether, "he seem'd not at all delighted with the thought, nor ever laid any stress upon it." Not so the new inquirers of the mid-century, for "we are arriv'd at great dexterity since Sir Isaac's time . . . [and] can now almost prove the existence of this aether by the phenomena of electricity." [13]

Here our concern is not with electrical experiments, but rather with the speculative theories of matter such experiments helped to provoke, and the influence of those theories on later chemical developments. For instance, one of the earliest theorists was the Dublin physician Bryan Robinson. His work was to influence the electrical researches of Benjamin Wilson and Benjamin Franklin. However, Robinson's ideas are of direct interest here because of their effect on the chemical thought of William Cullen, and their possible influence on the development of the caloric theory.

5.3.1. Bryan Robinson and the Medical Tradition

Robinson[14] is yet another member of the second-generation group of British Newtonians—the group that grew to maturity under the influence of the master and his immediate circle of disciples. Little is known of Robinson's early life, but already in 1704 he was dedicating a translation of de la Hire's *New Elements of Conick Sections* to his "best friend" John Harris, who, through the production of the *Lexicon Technicum*, had "done more service to true philosophy, and useful learning, than any one single person besides." Apparently it was Harris who suggested the translation, as well as "correcting the sheets,

---

[13] Martin (1747), *Philosophia Britannica*, I, sigs. a2^r and a3^v. The whole question of experimental work on static electricity in the 1740's and 50's is explored at length in Cohen (1956), *Franklin and Newton*.

[14] (1680–1754). See *DNB*.

in my [Robinson's] absence." [15] The translation bore a London imprint, but Robinson himself may well have been in Dublin by the time the work appeared.

We do know that Robinson chose to read medicine at Trinity College, Dublin, graduating M.B. in 1709 and M.D. in 1711, and that he then went on to practice in Dublin with great success for the rest of his life. Not surprisingly, he became a close friend of Richard Helsham,[16] another second-generation Newtonian and Trinity graduate. Helsham himself was successively fellow, senior fellow, first professor of natural philosophy (1724–1738), and Regius Professor of Physic (1733–1738) in the college. After his death, Robinson undertook the publication of Helsham's *Course of Lectures in Natural Philosophy*. Like the similar Newtonian texts of 'sGravesande, Desaguliers, and Rowning, this work enjoyed wide popularity, reaching a fifth London edition in 1777 and even earning a Dublin reprint of "Select parts" as late as 1818. In view of other information on Helsham and Robinson, discussed below, the main interest of the *Course of Lectures* lies in its total avoidance of the ether—an avoidance typical of Newtonian works of the period, but atypical of Robinson or, apparently, of Helsham.

Disappointingly little is known of Robinson's contact with either the original Newtonian circle or later workers in London. A passage in his 1752 *Observations on the Virtues and Operations of Medicines* does seem to imply a correspondence, or possibly a meeting, between Newton and Helsham, with the ether as the subject of discussion. The passage states:

> By the answer he [Newton] gave to Dr. Helsham, who asked him, how muscular motion is performed by the vibrations of the aether, it appears that he excluded all animal fluids . . . from having any share in causing this motion. The answer

[15] Robinson (1704), sigs. a 2^v–a 4^r.

[16] (1682?–1738). See *DNB*. Both Robinson and Helsham were really on the border line between the first generation of British Newtonians, who came to intellectual maturity in the 1690's, and those who grew up in a scientific world dominated by Newton.

was this, "I suppose, said he, the vibrations of the aether propagated from the brain through the nerves to the muscles to agitate their membranes, and to swell them without heat." [17]

It would be extremely interesting to know more of this interchange between Newton and Helsham. Its occurrence might well explain the prominent role given to the ether in Robinson's own 1732 *Treatise of the Animal Oeconomy*. This latter work was firmly inside the Newtonian medical tradition, Robinson choosing to praise especially Harvey, Lower, and "Sir Isaac Newton [who] discovered the causes of *muscular motion*, and *secretion;* and likewise furnished material for explaining *digestion, nutrition,* and *respiration.*" [18] That it was in a 1730's medical work that Robinson first developed his ideas on the importance of the Newtonian ether in the system of Nature again illustrates the importance of medicine to natural philosophy in general, and chemistry in particular, all through the eighteenth century.

The whole *Treatise* was set out in the form of propositions, corollaries, scholia, and "proofs by experiment," so great was Robinson's desire for a Newtonian medicine. The most interesting proposition for our purpose is number 8, which states that "muscular motion is performed by the vibrations of a very elastick aether, lodged in the nerves and membrances investing the minute fibres of the muscles, excited by the power of the will, heat, wounds, the subtile and active particles of bodies, and other causes." Newton's support for this position was adduced from the Queries in *Opticks,* quotations from which occupy two whole pages, and Robinson felt confident enough to assert that "since *the other phaenomena of Nature absolutely require such an elastick fluid,* as is the aether described by Sir Isaac Newton; and since causes are not to be multiply'd without

---

[17] Robinson (1752), 57. A further hint of Dublin-London contact occurs in a 1725 note that a tract on smallpox inoculation "by Dr Bryan Robinson" was read to the Royal Society: *Journal Book of the Royal Society (Copy),* XII, 586.

[18] Robinson (1732), iii.

Atoms and Powers

necessity. Therefore it must be granted, that this motion begun in the nerves at their origin, *is the vibrating motion of that aether*."[19] The implications of the position thus adopted by Robinson were to be fully spelled out a decade later in his 1743 *Dissertation on the Aether of Sir Isaac Newton*.

The opening paragraph of the *Dissertation* makes the work's purpose so clear that it is worth quotation in full:

> In this dissertation on the aether of Sir Isaac Newton [said Robinson], I shall first give an account of the nature of an elastic fluid; secondly, I shall shew how a subtile aether causes an universal attraction and gravity; thirdly how it causes repulsion and elasticity; fourthly, how it causes the principal phaenomena of light, namely, its emission, refraction, reflexion, inflexion, and fits of easy transmission and easy reflexion; fifthly, how it causes heat, and the rarefaction of bodies by heat; sixthly, how it causes muscular motion and sensation; and, seventhly and lastly, I shall shew how it makes the particles of bodies to cohere, and causes fermentation.[20]

The program was then faithfully followed out in the long series of propositions, corollaries, and scholia that made up the main bulk of the *Dissertation*.

Robinson's work provides the best example of many contemporary attempts to systematize the position at which Newton was continually hinting in his published writings, but the difficulties of which he never overcame. (No more did Robinson or other commentators but they were armed with faith in Newton, not Newton's caution.) Inert gross matter and mutually repelling ether particles formed the basis of Robinson's system. True to his theological heritage, he was careful to point out that spiritual powers were the ultimate causes, because the repulsive power itself was not material "and therefore the cause, which gives the aether its activity and power, must be spirit."[21] Robinson's adoption of a hierarchical internal structure to gross mat-

---

[19] *Ibid.*, 82 and 93. Italics are the present author's.
[20] Robinson (1743), 1–2.
[21] *Ibid.*, 122.

ter, and his differentiation between the observed force laws of gravity and those of fermentation and cohesion were thoroughly in keeping with orthodox Newtonianism. But while there was little new in his exposition, we cannot therefore neglect it.

For a start, it is tempting to think that it was Robinson's exposition—published in Dublin in 1743—which prompted Thomas Birch to include the *Letter to Boyle* (of which Robinson was apparently unaware) in his 1744 edition of Boyle's *Works*. This in turn provoked Robinson to publish his 1745 *Sir Isaac Newton's Account of the Aether, with some Additions by Way of Appendix*, almost half of which was taken up with a straight reproduction of the *Letter*. And the *Letter* certainly assisted that growth of imponderable-fluid theories which was so marked over the next quarter-century. Then again, Robinson's work greatly encouraged Benjamin Wilson in his electrical experimenting and theorizing, Wilson even traveling to Dublin in 1746 to see Robinson and work with him.[22] One result seems to have been an *Appendix* to the *Dissertation* (dated 1 December 1746) in which Robinson discussed electricity, while another was Wilson's own *Essay towards an Explication of the Phaenomena of Electricity, deduced from the Aether of Sir Isaac Newton*. A third was the appearance of a 1747 London edition of the *Dissertation*, soon after Wilson's return from Ireland. Wilson's work was itself vital to the formulation of Franklin's one-fluid theory of electricity, and the success of that particular imponderable-fluid theory could not but influence the course of later chemical thought on phlogiston and caloric. Finally we may note that the appearance of the London edition of Robinson's *Dissertation* was probably the immediate cause of the development of Gowin Knight's rival scheme. But for an example of the direct impact of Robinson's work on chemical theory we cannot do better than turn to the lectures of William Cullen.[23]

---

[22] Wilson (1746), xii. See also Cohen (1956), *Franklin and Newton*, 417–424 on Wilson's work, and Turner (1967), "James Short, F.R.S.," for some nonscientific reasons for Wilson's travels.

[23] (1710–1790). See Thomson (1832), *Life of Cullen*, and Wightman (1955–56), "Cullen and chemistry."

Atoms and Powers

First under Cullen, then under his pupil Joseph Black, the medical and chemical teaching of Glasgow and Edinburgh Universities in the later eighteenth century came to play much the same role on the European scene as the teaching at Leyden under Herman Boerhaave at an earlier period. It is therefore of considerable significance to the development of chemical thought that, though he believed chemistry "not yet ripe to be reduced to a complete system," William Cullen was declaring in his 1762/63 lectures that "Dr. Bryan Robinson's of Dublin, treatise on the aether of Sir Isaac Newton . . . is the only probable scheme of a chemical theory." Cullen further asserted that "all the operations of Nature and of Art depend upon attraction and repulsion" and that

> by thus observing the various states of the aether in bodies we shall understand the forms of solid and fluid, the various states of concretion, & consequently the various properties depending on those states. Or if the phenomena of Nature cannot be found under these heads they will be understood from the consideration of the motions derived from the repulsive powers, which occur in the transposition of bodies.

He also pointed out that though "the doctrine of elective attraction presents us with some difficulties when we adopt the hypothesis of the aetherial fluid," yet "if we consider the analogy of the electric fluid, this difficulty will vanish." [24]

Cullen's lectures show with great clarity how by the 1760's Newtonian ideas on matter, and the associated beliefs in internal structure, short-range forces, and an elastic-fluid ether, had become fundamental to the physicalist-reductionist school of chemistry. It was this school and consequently this view of mat-

---

[24] Quotation from Manchester University Library, Special Collections, Ms. no. CH C 121 1, pp. 43–45. On Joseph Black (1728–1799), see Ramsay (1918), *Life of Black,* and the series of articles by McKie (1936–66). Before his death in 1967, Professor McKie had been working to produce an adequate biography of Black, and thus fill in a major lacuna in the history of eighteenth-century natural philosophy.

ter which was to dominate the chemistry of the latter part of the century. Before going on to consider in detail some aspects of this domination, we must deal with developments which finally modified Newtonian matter-theory beyond hope of Newton's recognition, if not that of his disciples.

5.3.2. Gowin Knight: Attractive and Repulsive Matter

Gowin Knight [25] is now usually known only as a somewhat obscure figure in the history of magnetism, but he was not without fame in his own day. Educated at Magdalen College, Oxford, he later settled in London as a physician. However it was his work on magnetism, not medicine, that brought scientific reputation and pecuniary reward. Discovering a means of making magnets more powerful than had previously been possible, he contributed a series of papers to the *Philosophical Transactions*. His work was rewarded by a Copley medal in 1747. The impression Knight's discovery created may be seen in a 1745 letter of Peter Collinson, the Quaker intelligencer, about "Dr. night a physition [who] has found the art of giveing such a magnetic power to steel that the poor old loadstone is putt quite out of countenance."

Encouraged by the reception of his work, and no doubt influenced by the appearance of Robinson's *Dissertation,* Knight composed his own speculative treatise (as did such other figures as Benjamin Wilson and Cadwallader Colden, Collinson's correspondent).[26] Published in 1748, Knight's work was wordily but accurately entitled *An Attempt to Demonstrate, that all the Phaenomena in Nature may be Explained by Two Simple Active Principles, Attraction and Repulsion: Wherein the Attractions of Cohesion, Gravity, and Magnetism are Shewn to be One and the Same; and the Phaenomena of the Latter are more Particularly Explained.* The *Attempt*'s importance is threefold. It

---

[25] (1713–1772). See Nichols (1817–58), *Literary History,* VIII, 626.

[26] See Wilson (1746) and Colden (1745). Such treatises enjoyed a considerable vogue around mid-century: for Buffon's favorable reaction to Colden's ideas, see Babson (1950), *Newton Bibliography,* 27. Peter Collinson's remark on Knight is quoted from Colden (1917–23), *Letters,* III, 114.

represents a conscious effort not only to systematize but also to modify and develop Newtonian matter-theory. Secondly it makes a definite, if unwitting, further step along the road toward the "immateriality of matter" and the haunting ideas of Boscovich. And, thirdly, it provides the first clear exposition of the ideas which later underlay the caloric theory of heat.

The *Attempt* made quite plain the importance of Newton's work, for "a greater progress has been made in physicks by Sir *Isaac Newton* alone, than by all the philosophers before or since him put together." Knight, like Newton, adopted the view that extension, impenetrability, mobility and the *vis inertiae* were general properties of matter with their immediate cause in the will of God. Displaying none of Newton's qualms, he also held that attraction and repulsion were universal causes, and as such equally ascribable to the will of God. In this Knight characterizes the change from the early years of the century, with its debates about the "occult" nature of gravity, to the later years in which "attraction" was to be an all-sufficient cause, theologically explicable by Englishmen, if only accepted with suitable positivistic agnosticism on much of the Continent.

More radical was Knight's realization that the traditional concept of atoms differing in shape and size was redundant in a world to which forces were admitted. Newton had supposed his primitive particles to be of varied configuration, though composed of the same primary matter. These particles were then ordered into complex hierarchies up to particles of the last composition, on which the normal operations of chemistry depended, and which might perhaps be observable with powerful microscopes. But for Knight "all the primary particles of matter are originally of the same size, and all round. This is more simple, than if we were to suppose them of different sizes and shapes, or of any other shape beside this. So that unless different sizes and other shapes were necessary, which as yet we have no reason to suppose, the truth of this proposition is an evident consequence of the last."

Knight's world did admit of some variety for, before arriving

Speculative Systems

at the above proposition, he had already decided that "attraction and repulsion cannot both, at the same time, belong to the same individual substance, being contraries ... Therefore we must conclude, that there are in Nature two kinds of matter, one attracting, the other repelling." His primary particles were therefore of two types—those which attracted each other and those which repelled each other, *but yet were attracted by the attracting particles.* In this he again advanced beyond the orthodox Newtonian view (as expressed, say, by Bryan Robinson) of inert gross matter and mutually repelling ether particles. In accord with his belief in simplicity, he also assumed that his attractive and repulsive forces both obeyed a $1/r$ law. Support for this assumption was found in such things as magnetism and air pressure, Knight deciding that it should be possible to reconcile other phenomena with the law.[27]

Though Knight sought to modify and extend Newtonian concepts, this was only because he implicitly accepted the Newtonian world-view. Thus his primary particles were to be "extremely minute" for, like Newton, he wished to build up a world of corpuscles possessing an internal structure. Further, in his emphasis on the infinite divisibility and extreme tenuity of matter, we see a clear acceptance of the "nut-shell" scheme popularized in Keill's *Introductio*. On the operational level, Knight concludes that his primary particles give rise to three types of complex corpuscle and that "all the various degrees of cohesion, that are found amongst bodies, will naturally arise from the different sizes of their constituent corpuscles; and one simple, uniform kind of attraction and repulsion, such as hath been already explained." In true Newtonian fashion it is these complex corpuscles that represent the chemical elements. Repellent corpuscles give "pure elementary air," while neutral corpuscles form "pure elementary water." Still adhering to the Empedoclean scheme, but influenced by Stahl, Knight asserts that the smallest attracting corpuscles will form "what *Stahl* and

---

[27] Quotations are all from the second edition: Knight (1754), sig. B 1ᵛ, and props. 7, 10, 14, 15, 19, and 24.

other chemists have called the *phlogiston*, the most subtle earth in Nature." Larger attracting corpuscles compose "the universal acid, the basis of all salts," and still larger corpuscles form stones, metals etc. Interesting as these passages are in throwing light on the chemical beliefs of the period, their major importance is in the way they illustrate the conviction that all observable chemicals are complex corpuscles, built out of primary particles.

Unlike Robinson, Knight was not content merely to systematize the Newtonian scheme. Instead he altered and went beyond it where this seemed either necessary or desirable. Hence the variety of traditional atomism was abandoned, as offering no advantage over entirely uniform ultimate particles. The enduring belief in the unity of matter was also modified, or at any rate questioned, in the assumption of primary particles of the same size and shape, yet of different type, some attracting, some repelling. Another important modification was that, though Knight accepted the porosity of all natural bodies in a way which directly echoed Newton in the *Opticks*, he abandoned the idea that the particles of bodies actually touch one another.

Instead, Knight said:

> All bodies whatsoever, whether solid or fluid, must contain more pores than solid parts. The truth of this proposition has been sufficiently proved by most of the philosophers of this and the last century, from facts and experience: yet it was not so easy to conceive, how it could be true in all cases; whilst they supposed the constituent parts of bodies in actual contact. But since it is evident that the corpuscles, of which bodies are compounded, never come in actual contact at all, I think all those difficulties are removed, and the truth too clear to require any further demonstration.

A natural corollary of this view was the concept of a limit point of equilibrium between two corpuscles:

Speculative Systems

Attracting corpuscles approach each other with an increasing force, till they come so near, that the repulsion at their surfaces begins to counter-act their force of attraction ... and at a certain distance the repulsion at their surfaces becomes so strong, as to equal and quite destroy the attracting force ... This law of cohesion very well explains the cause of elasticity in bodies.[28]

In these statements we again see a "logic of development" at work in Newtonian commentators. Newton's own writings had left a number of ambiguities unresolved, a number of lines of development unpursued. While his immediate disciples might recognize the problems, and respect the ambiguities, later commentators felt no such needs. The molding of a coherent body of natural philosophy, capable of widespread diffusion and textbook transmission, demanded that uncertainties be removed—and therefore that lines of development be pursued. Rowning was already groping toward the concept of limit points of equilibrium in a 1730's text. Knight's *Attempt* merely took the next step. And a decade later, Boscovich's far more powerful intellect was to pursue the issue to its logical conclusion.

The desire to systematize Newton was beset by theological problems throughout the eighteenth century. That questions about the structure and properties of matter should directly concern both believing Christians and believing pagans was natural in a world where natural philosophy and religion were not divorced, and amateur status was still the norm for men of science. Newton's own writings bear eloquent testimony not only to his religious concerns but also to the theological context of his continued tension over alternative modes of explanation. Forces such as those of gravity, fermentation, and cohesion could be treated simply as known properties of matter. As such they were either essential and inherent to matter (as Continental positivists would hold) or merely observed physical results of Divine and immaterial action (as Newton and English

[28] *Ibid.*, props. 48, 50, and 49.

deists preferred). Alternatively the whole problem might be avoided, and all forces "explained" by reference to some single physical cause—in Newton's case a repulsive-fluid ether.

Newton hesitated to embrace the ethereal alternative. The reasons perhaps include his earlier opposition to Cartesian and Leibnizian versions of the plenum, his awareness of the conceptual problems associated with nongravitating matter, and his own satisfaction with the idea of forces as the direct manifestation of God's omnipresence in Nature. After Newton's death, this last position was gradually and subtly transformed until it became less a defense against, than a bulwark of, materialistic and rationalistic systems. If matter, beside extension, divisibility, mobility, and the *vis inertiae*, might also be taken to possess (for purposes of physical argument) such forces as those of gravity, etc., why might not thought and life themselves be merely a function of the complex organization of this remarkable—and still largely unknown—thing called matter?

Such conclusions might delight radical Christians like Joseph Priestley—but not the theologically conservative, among whom must be numbered not only Boscovich, but also the high Anglicans, and the whole burgeoning English Evangelical movement. The deep concern these latter groups felt to maintain the unbridgeable divide between matter and spirit led to strange results, most notable in the systematics of the Hutchinsonians. The need to emphasize the inertness, grossness, and passivity of brute matter became increasingly urgent as the century progressed. The challenge of Priestley's 1777 *Disquisitions on Matter and Spirit* was to unite a variety of conservative traditions in English thought, embracing not only eminent Hutchinsonians such as Bishop Horne, and prominent Evangelicals like John Wesley, but also such Quaker figures as John Whitehead and John Dalton. Paradoxically, it was to be these exalters of the spirit who were thus more and more impelled to stress the earthy materiality of matter, and the existence of ether-like mechanisms as God's chosen means of operation in Nature. We shall refer to this paradox again in Chapters 6 and 8, but now we

Speculative Systems

must return to the more immediately physical and directly scientific.

The 1740's and 1750's proved to be the great age for English-language systematizations of Newton's work. A generation later, natural philosophers were neither so sanguine about all-inclusive schemes nor so enthused of the virtues of one universal, weightless, repulsive-force, fluid medium. On the one hand, the systematizers had failed to produce much apart from words. On the other, the phenomena of heat, light, combustion, magnetism, electricity, and nervous impulse were becoming increasingly complex and recalcitrant with each fresh experimental discovery. All-embracing schemes were at a discount. Specialist studies and limited hypotheses were preferred. Typical of both was William Cleghorn's treatise on fire—a treatise which demonstrates the considerable explanatory power repulsive-force theories still possessed, even when constrained within narrow limits.

5.3.3. Cleghorn and Caloric

The fact that William Cleghorn [29] spent his undergraduate days at Trinity College, Dublin, may help to explain his later attempt to construct a more modest version of Bryan Robinson's *Dissertation*. Exchanging Dublin for Edinburgh and the study of medicine, Cleghorn naturally attended the classes of Cullen and Black. The latter's work on heat seems to have been the immediate cause of the *Disputatio Physica Inauguralis, Theorem Ignis Complectens*, or *De Igne*, which was both Cleghorn's 1779 M.D. dissertation and a speculative treatise that tells us much about the mood and style of Edinburgh chemistry at the time.

According to the posthumously prepared and heavily edited version of his *Lectures on Chemistry*, Joseph Black was averse to all speculation which required him to think "not as a chemist, but as a mechanician." In keeping with such a judgment, the edited *Lectures* tell us how Black read "Dr Gowen Knight's

---

[29] (1754–1783). See Heathcote and McKie (1958), "Cleghorn's *De Igne.*"

ingenious essay on magnetism" with disapproval, for "the first steps of the investigation seemed to give very clear notions; but he [Black] found them afterwards involved in continual inconsistencies." Considering the encouragement he gave to Cleghorn's work, and what little else we know of Black's private thoughts, this statement seems rather a tribute to the prejudices of editors than a reliable piece of historical information.

Black's enthusiasm for Cleghorn's work has an additional interest. He seemed to think the latter's chief merit and originality lay in expounding just that view of matter put forward by Gowin Knight more than thirty years earlier—so little were completely general schemes such as Knight's acknowledged in the 1770's. Black said:

> He [Cleghorn] supposes, that heat depends on the abundance of that subtile fluid elastic matter, which had been imagined before by other philosophers to be present in every part of the universe, and to be the cause of heat. But these other philosophers had assumed, or supposed one property only belonging to this subtile matter, vis. its great elasticity, or the strong repellency of its particles for one another; whereas Dr Cleghorn supposed it possessed another property also, that is, a strong attraction for the particles of the other kinds of matter in Nature, which have in general more or less attraction for one another. He supposes, that the common grosser kinds of matter consist of attracting particles, or particles which have a strong attraction for one another, and for the matter of heat; while the subtile elastic matter of heat is self-repelling matter, the particles of which have a strong repulsion for one another, while they are attracted by the other kinds of matter, and that with different degrees of force.[30]

As this exposition makes clear, Cleghorn used the same basic principles as Knight. However there were important differences of emphasis between the two men. While Knight desired to

---

[30] Black (1803), I, 516–517 and 33–34. For evidence of how severe was the editing of the published *Lectures*, see McKie (1960), "Letters of Black."

Speculative Systems

frame an explanatory scheme embracing all Nature, Cleghorn's concern was at once narrower and more empirically based.

Boerhaave had supposed the capacity for heat of any substance to be a direct function of its inertial mass. It was the failure of this idea to account for observed temperature changes that led to Black's work on specific and latent heats.[31] It is thus not surprising that Cleghorn, as a pupil of Black, should, after reviewing his master's experiments, conclude that "there is in bodies a force attracting fire and . . . this is different in different bodies" and that "fire is distributed among bodies directly in proportion to the power by which they attract fire and inversely . . . to the repulsive power of the particles of fire."[32]

Cleghorn made no attempt to explain how this attraction of fire was related to the internal structure of the bodies concerned, nor why particles of fire repelled one another, nor what power law was involved in these attractions and repulsions. In this way the theory reflected the continuing heritage from mid-century speculative attempts to "explain all the phenomena of Nature" with the aid of mutually repelling particles. But unlike Newton's ether and Knight's repelling matter, Cleghorn's fluid was not a universal explanation principle, but rather fire itself, and as such a material distinct from things like phlogiston and electricity. This does not mean that Cleghorn, Black, or other such "physicalist" inquirers into chemistry, physiology, and medicine saw these fluids as necessarily fundamental and incapable of reduction to one all-pervasive ether. Rather they saw them as possessing immediate explanatory utility, and so were willing to employ them, without prejudice to the larger issues.[33]

Cleghorn's theory was more detailed, and more related to particular experimental results, than those of either Knight or Boscovich. Yet Cleghorn's work was no more predictive than his

[31] See Heathcote and McKie (1935), *Specific and Latent Heats*. See also Thackray (1968b), "Matter in a nut-shell," note 28, for a 1720 Royal Society experiment based on Boerhaave's supposition.

[32] Heathcote and McKie (1958), "Cleghorn's *De Igne*," 21.

[33] In a lecture at Harvard University in November 1967, Professor R. E. Schofield argued that British natural philosophers of the mid-eighteenth century can best be viewed as either "mechanists" or "materialists." A full appraisal must await Professor Schofield's promised book-length study of the subject.

predecessors' for, like theirs, it could be reconciled with widely varying experimental results. Its greater usefulness lay in its provision of a more adequate conceptual model for fire, through its concentration on this alone. Further modified to give a theory of caloric or heat, the sort of concepts that Cleghorn employed were to become integral to Lavoisier's revised chemistry. Their continuing role in the new chemistry of the early nineteenth century is at once clear from the opening words of Dalton's *New System of Chemical Philosophy*: "The most probable opinion concerning the nature of caloric, is, that of its being an elastic fluid of great subtilty, the particles of which repel one another, but are attracted by all other bodies." [34]

We cannot follow the later, and increasingly involved, history of the imponderable fluids. As ever more complex experimental phenomena were encountered, magnetism, electricity, and nervous impulse each came to demand their own fluid, beside the fluid of heat and the original gravity-causing ether. Professor Cohen's exemplary study of theories of the electric fluid shows how much there still is to learn about the subtleties and interrelations of this aspect of eighteenth-century thought. Clearly, experimental and theoretical developments in electricity influenced thought and research on, say, phlogiston and light. But until further monographic studies have been undertaken, it is impossible to say anything more specific, and therefore relevant to this present essay. We shall instead turn from fluid theories and Cleghorn's work to the earlier and far different system of R. J. Boscovich.

5.4. R. J. Boscovich and the Theoria Philosophiae Naturalis

Though by the end of the century most British natural philosophers had abandoned the Newtonian enterprise of providing

---

[34] Dalton (1808–27), I, 1. There is no evidence that Lavoisier knew of Cleghorn's treatise. His ideas seem rather to have derived largely from Boerhaave's work. What is important here is that imponderable-fluid theories of heat enjoyed unequivocal Newtonian legitimation in both Scotland and France. On caloric theory in early-nineteenth-century France, see Fox (1968), "Background to Dulong and Petit."

an all-embracing "theory of natural philosophy," many of them were deeply influenced by the last thoroughgoing eighteenth-century attempt to do just that. Reference to the ideas of R. J. Boscovich [35] may be found in the writings of such varied authors as William Cleghorn, Joseph Priestley, John Michell, William Herschel, John Robison, Dugald Stewart, Thomas Thomson, Humphry Davy, John Dalton, Michael Faraday, and William Hamilton. Among this group, men with chemical interests are prominent. Their knowledge of Boscovich's ideas was to have important effects on the course of chemical debate.

By 1727 Robert Green had realized that to admit force as an explanatory concept was to render obsolete the defining of matter in terms of solidity and extension. In 1748 Gowin Knight demonstrated the redundancy of Newton's assumption of varied shape among his primitive particles. Though apparently unaware of both these authors, and himself within a very different tradition, Boscovich was to combine their ideas in his sophisticated attempt to recast the way that natural philosophers regarded the world.

His *Theoria*, or *Theory of Natural Philosophy, Reduced to a Single Law of the Forces Existing in Nature,* was published in Vienna in 1758. It stands supreme among eighteenth-century speculative treatises, as much for the clarity of its arguments as for the power of its fundamental assumptions. Boscovich's *Theoria* also stands quite apart from the British treatises we have been discussing in the motives that lay behind the work. It was the author's special desire to heal the bitter breaches between Newtonians and Leibnizians by production of a system acceptable to both. In particular Boscovich owed to Leibniz the idea that the ultimate material elements were dimensionless points, and also the law of continuity which provided an essential basis for his arguments. For the idea of interparticle attractive and repulsive forces, he was indebted to Newton and to Query 31 of *Opticks*.

[35] (1711–1787). See Whyte (1961), *Boscovich Studies,* for a collection of essays expressing the newer interest in this urbane and talented Jesuit.

Making quite explicit its aim, the *Theoria* begins: "The following theory of mutual forces ... presents a system that is midway between that of Leibniz and that of Newton ... and, as it is immensely more simple than either, it is undoubtedly suitable in a marvellous degree for deriving all the general properties of bodies ... by means of the more rigorous demonstrations."[36] The essence of the theory is given in the synopsis, where Boscovich says that "matter is unchangeable, and consists of points that are perfectly simple, indivisible, or no extent, and separated from one another; that each of these points has a property of inertia, and in addition a mutual active force depending on the distance." Boscovich accepts the fundamental unity and inertial homogeneity of all matter, arguing for it both from the trend revealed by chemical analysis, and from the alphabet analogy. However the ultimate units of matter are not only homogeneous, as in Newton, and all alike, as in Knight, but also entirely dimensionless. These dimensionless "points" possess only inertia and a property of relative acceleration when placed in relationship to each other. The number of explanatory categories is therefore reduced to two—force, or more strictly length, and time.

As with Descartes's very different earlier attempt to reduce everything to two categories, so with this reduction a price was paid. In this case it was the introduction of a force-curve of highly arbitrary shape (see Fig. VII): a step which, though philosophically dubious, appealed to an age intent on discovering and measuring those forces between particles that Newton's speculations had revealed. Besides the abandonment of size, a further break with the orthodox Newtonian position was the conclusion that physical contact between any two particles was impossible. At this point Boscovich developed the idea, present in Gowin Knight's work, that at infinitely small distances the force-curve tended to infinite repulsion, not infinite attraction,

---

[36] Boscovich (1763), par. 1. All quotations are by numbered paragraph from the 1922 English edition prepared from the 1763 Venice edition. This latter is usually taken as authoritative, having been published under the author's direct supervision, unlike the 1758 edition.

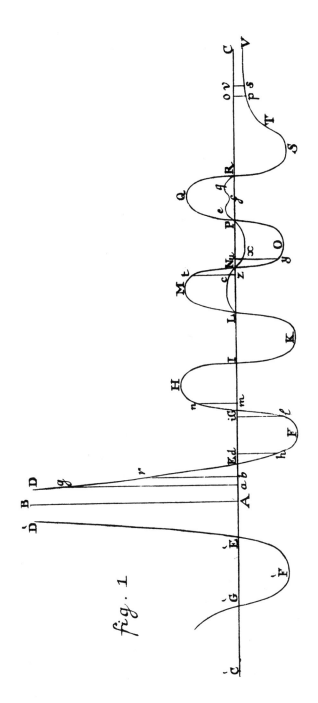

VII. Boscovich's force-curve. From Boscovich (1763), plate I.

as Newton had supposed and John Keill elaborated. He also sided firmly with the nonethereal interpreters of Newton in supposing that interparticle forces simply alternated from attraction to repulsion as distance varied, without any need to suppose two fundamentally different types of matter.

In the *Theoria,* Boscovich brought out the logical implications of his ideas with great skill. He showed how from the interaction of three or more points, located in three dimensions, stable systems or "particles of the first order" might be obtained. From these, "particles of the second order" resulted "and so on; until at last we reach those . . . variable particles, which are concerned in chemical operations . . . with regard to which we get the very thing set forth by Newton, in his last question in Optics, with respect to his primary elemental particles, that form other particles of different orders." Typical of the depth and sophistication of Boscovich's thinking was the way in which he conducted an analysis of the interactions of complex particles. He showed how in the extreme case "the same particle in one part may exert attraction on another particle, and repulsion from another part; indeed, there may be any number of places in the surface of even a spherical particle, which attract another particle . . . whilst others repel, and others have no action at all." [37] In this way he deductively developed a picture of "directed valencies" between particles. Small wonder that his ideas were to fascinate such able chemists as Humphry Davy and Thomas Thomson.

The great bulk of the *Theoria* was occupied with a general exposition of Boscovich's ideas. The aim throughout was not to predict, but to reconcile the theory with all the varied phenomena of experience. Like other writers of speculative treatises from Descartes to Gowin Knight, Boscovich was concerned not with predictive, and hence falsifiable, theories. Instead his aim was that of the true eighteenth-century natural philosopher: providing explanations which satisfied the intellect by a demonstration of the simplicity underlying all the variety of Nature. His last hundred or so paragraphs did deal with chemical phe-

---

[37] Boscovich (1922), pars. 239 and 423.

nomena, such as solidity, fluidity, precipitation, solution, crystallization, light, and fire. But he was more than content to remark that "I really think that there can be found in this *Theoria* the general theory for all chemical operations," and to admit how "the special determination of effects that arise from each of the different mixtures of the different bodies, through which alone all effects in chemistry are produced . . . would require . . . the whole power of geometry and analysis, such as exceeds by far the capacity of the human mind." [38]

Its lack of "predictive power" should not lead us to ignore the *Theoria,* which exercised a powerful influence on later thinkers. Admittedly, it had little to offer imponderable-fluid theorists, but (a situation Boscovich did not intend) its categories had considerable appeal to British radical thinkers. On a purely scientific level, the *Theoria* was important for its stress on attractive and repulsive forces, and on the significance of the internal structure of matter. Just as in Keill, Freind, and Buffon, so in Boscovich there was a strong belief that chemical understanding must be sought through the study of forces. However it is worth remarking that Boscovich's system, unlike Buffon's, held out little incentive to undertake practical, experimental quantification of chemical forces. The interaction of so many force-curves would after all create a situation "such as exceeds by far the capacity of the human mind." To followers of Boscovich, experimental measurements of observed forces could offer little information of any significance—a fact that helps to explain the paucity of British attempts at Newtonian chemical quantification. Certainly, the most active experimenters in the field seem to have been inspired by the rival Buffonian vision of the universal rule of a $1/r^2$ law.

5.5. Clairaut, Buffon, and the Law of Attraction

George-Louis Leclerc, Comte de Buffon,[39] is best known as the author of the mammoth *Histoire Naturelle*. But, as a recent

---

[38] *Ibid.,* par. 451.
[39] (1707–1778). See Buffon (1860), *Correspondance,* and (1863), *Buffon.*

study has vividly demonstrated, his education and early interests were in mathematics and Newtonian physics, not the biological pursuits which later brought him renown. Already in 1730, at the age of twenty-three, Buffon was engaged in European travels with Nathaniel Hickman, a Leyden-educated fellow of the Royal Society. Arriving in Paris in July 1732, he was soon acquainted with the moving spirits in the endeavor to subvert the Cartesian orthodoxy of the Académie Royale des Sciences. Significantly, it was to be a favorable report by two leaders of the subversive movement, Maupertuis and Clairaut, which gained Buffon admission to that same Académie, early in 1734.[40]

If anyone in France had any doubts as to Buffon's Newtonian loyalties, his 1735 translation of Hales's *Vegetable Staticks* was enough to lay them to rest, to say nothing of his 1740 French version of Newton's *Treatise on Fluxions*. In this latter, Buffon came out firmly on Newton's side in the calculus-priority dispute: so much was acceptance of Newtonianism still associated with faith in Newton and repetition of anti-Leibniz shibboleths. Buffon's energy and enthusiasm, his dedication to Newtonian tasks, and his continuing intimacy with Maupertuis and Clairaut, help to explain why Voltaire was facetiously to claim in 1739 (apropos of his own Newtonian activities) that "I am the spoiled child in a party of which Buffon is the head."[41]

Buffon's own predilection for systematics may be glimpsed in his preface to the translation of *Vegetable Staticks* where, despite many protestations of the necessity for facts, experiments, and experience, he also "asserts that nothing would be so good as first establishing a single principle, from which to explain the universe; and I conclude that were one fortunate enough to divine it, all the trouble one undergoes in making experiments would be useless."[42] In their context, the words

---

[40] See the detailed study by Hanks (1966), *Buffon avant* "l'Histoire Naturelle," which the present author has drawn on freely. See also Brunet (1936), "Buffon—disciple de Newton," and Wohl (1960), "Buffon's project for a new science."
[41] Quoted from a letter of 3 October 1739, in Hanks (1966), 91.
[42] Buffon (1735), iv.

Speculative Systems

were intended to be ironic. In view of Buffon's later statements, the irony seems at best two-edged, for it was just such a taste for "single principles" that led first to a sharp exchange within the French Newtonian camp and, subsequently, to Buffon's all-embracing system.

The trouble began in 1747, with a paper delivered to the Académie des Sciences by the mathematician A. C. Clairaut.[43] Like Buffon, Clairaut was a convinced Newtonian. His credentials included collaboration and friendship with Maupertuis, instruction to Madame du Châtelet, Newtonian participation in French studies on the shape of the earth, and writings in defense of Newton's optical theories. The public argument that broke out between the two men thus reflects both the security which Newtonianism enjoyed in France by 1747, and the ambiguities to which Newton's work so easily gave rise.

In his paper Clairaut provided a solution of the "three body problem." The paper also gave an exposition and defense of the Newtonian system of the world. However, Clairaut had to admit that, to his astonishment, his calculations showed the period of the apogee of the moon to be twice the observed value. In order to reconcile theory and experiment he suggested that the gravitational force law should contain another term. In this way a $1/r^2$ force at long distances could be reconciled with a higher power law at shorter (astronomical) distances of the order earth-to-moon. And, as he pointed out, such an amended law would also embrace "phenomena which are right before our eyes, such as the roundness of drops of fluid, the ascension ... of liquids in capillary tubes ... etc." [44] Less than three months later Buffon was to counterattack in the name of Newtonian orthodoxy.

Buffon did not attempt to deny the validity of Clairaut's calculations, but rather asserted that an apparent anomaly in the moon's motion was no reason for altering the law of gravitation. He vigorously opposed the new formulation, principally on the

---

[43] (1713–1765). See Brunet (1952), *Vie et Oeuvre de Clairaut*.
[44] Clairaut (1745a), 338.

ground that it "destroys the unity of law on which the truth and beautiful simplicity of Newton's system is founded." [45] In reply Clairaut pointed out that there did in fact exist a great number of phenomena which followed laws other than that of the inverse square. Undaunted, Buffon rushed in with a further paper "to demonstrate that the law of attraction in relation to distance can only be explained by a single term." [46]

In May 1749 Clairaut found that, by further calculation, the lunar apogee was reconcilable with a $1/r^2$ law. Even so he could not resist another round to the polemic. No more could Buffon resist replying that "metaphysical, mathematical and physical reasons all agree in proving that the law of attraction can only be explained by a single term." To which Clairaut's parting shot could only be a firm reiteration that "the metaphysical, mathematical and physical reasons that Mr. Buffon has advanced are of no effect against the law that I propose to reconcile astronomical phenomena with those that take place before our eyes every day, such as the roundness of drops of water, the ascension of liquids in capillary tubes, etc." [47]

This inconclusive exchange is of interest not only for the way it illustrates the tension between a desire to systematize Newtonian ideas and a concern for the observed effects. It is also important as a key to the understanding of Buffon's later insistence that even chemical attraction should be explained by a $1/r^2$ law. This insistence marked a radical departure from the whole previous style in physicalist and reductionist discussion of how chemistry was to be brought within the jurisdiction of Newtonian law. Perhaps because the departure was radical, and certainly because of his own later powers of patronage, Buffon's ideas deeply influenced the evolution of French chemistry.

The extended, formal exposition of Buffon's belief in the applicability to chemical investigation of the universal $1/r^2$ attractive law finally appeared in 1765, in volume XIII of the

[45] Buffon (1745b), 496 and 499.
[46] Clairaut (1745b), 531, and Buffon (1745b), 551.
[47] *Mémoires de l'Académie Royale des Sciences* (1745; published 1749), 577–586.

Speculative Systems

Histoire Naturelle. The volume was prefaced by a section which clearly displayed Buffon's love of systematization. It was in this section (entitled "Second view of Nature") that he discussed chemistry. His remarks show how different a guise the mid-century desire to "tidy up" Newton's work took in a country where the dismissal of final causes, the acceptance of attraction as essential to matter, the encouragement of rigorous analytical mathematics and positivistic approaches to natural philosophy, all went hand-in-hand with hostility toward revealed religion.

The chemical part of Buffon's discussion runs as follows:

> The laws of affinity . . . are the same with that general law by which the celestial bodies act upon one another. Their exertions are mutual, and proportioned to their masses and distances. Globules of water, of sand, or of metal, act upon each other in the same manner as the earth acts upon the moon: And, if these laws of affinity have hitherto been regarded as different from those of gravity, it must be ascribed to the confined views we have taken of the subject. Figure, which, in the celestial bodies, has almost no effect upon their mutual action, because the distance is immense, has great influence when the distance is very small . . . *All matter is attracted in the inverse ratio of the square of the distance; and this law seems to admit of no variation in particular attractions, but what arises from the figure of the constituent particles of each substance; because this figure enters as an element or principle into the distance.* Hence, when they [Posterity] discover, by reiterated experiments, the law of attraction in any particular substance, they may find, by calculation, the figure of its constituent particles. To make this matter more clear, let us suppose, that, by placing mercury on a perfectly polished surface, we [found], by experiment . . . that the attraction of mercury was in the inverse ratio of the square of the distance, it would be demonstrated that its constituent particles are spherical; because a sphere is the only figure which observes this law, and, at whatever distance

globes are placed, the law of their attraction is always the same.

Newton conjectured, that chemical affinities, which are nothing but the particular attractions we have mentioned, were produced by laws similar to those of gravitation. But he seems not to have perceived, that all these particular laws were only simple modifications of the general law, and that they appeared to be different, only because, at very small distances, the figure of atoms which attract each other has a greater influence upon the expression of this law, than the mass of matter.

Upon this theory, however, the intimate knowledge of the composition of brute matter solely depends.[48]

The logical simplicity and powerful appeal of Buffon's position is immediately apparent. So is the intellectual lure of making chemical reaction, and all other short-range phenomena, obey the $1/r^2$ law of gravitational attraction. The stress on measurement and calculation was to find a ready response among Continental chemists anxious to emulate the mathematical triumphs of Newtonian astronomers and physicists. In its possession of this stress Buffon's work differed markedly from that of the British systematizers, as also from that of the greatest systematist of all, R. J. Boscovich. Some of the factors underlying this profound difference in style and content have been hinted at, but any full exploration would take us far from present themes. Instead we may just note how the extrascientific reasons for Boscovich's appeal in Britain, and Buffon's in France, may well explain later differences in emphasis among Newtonian chemists in the two countries. To consider these differences in more detail is the task of the next two chapters.

[48] Buffon (1749–67), XIII, xii–xiv. Translation quoted from Buffon (1785), VII, 100–102.

6
The Problem of the Elements

> The Newtonian doctrine of elements is in the singular number: One sort of matter only makes the substance of all the infinite variety of bodies we behold.
> *Benjamin Martin in 1747.*

6.1. Introduction

Newtonian matter-theory may be seen as containing the three separate strands of belief in short- and long-range forces, belief in inertially homogeneous matter with a complex internal structure and, sometimes, belief in the existence of an imponderable, repulsive-fluid ether. The chemical implications of the first two beliefs are easier to handle and more directly relevant to this present study. This is partly because any treatment of the third raises the vital, but vast and vexing, question of eighteenth-century thought on heat, light, air, combustion, and respiration. It is also partly because imponderable fluids began to multiply at an alarming rate in the latter part of the century. And, not least, it is because imponderable fluids (in the guise of caloric) were the one part of Newtonian matter-theory to survive as an essential part of the new chemistry of the early nineteenth century.

The other two strands in the Newtonian view of matter were not so fortunate. Belief in inertially homogeneous matter with a complex internal structure was increasingly difficult to reconcile with the stubborn variety of chemical experience. Whatever Newtonian theory might demand, the presumed divorce between unobservable, homogeneous physical atoms and known, heterogeneous chemical elements became more difficult to defend. And the hope of quantifying the forces underlying chemical *mechanism* remained obstinately unfulfilled. Newtonian theory might dictate the existence of short-range forces on which the phenomena of chemistry depended: it could not dictate their successful measurement.

We shall not pursue the development of imponderable-fluid theories. Rather than discussing the one strand of Newtonian thought that survived, it seems more fruitful to concentrate on the endeavor to measure short-range forces, and the belief in homogeneous matter. These were the twin areas in which Newtonian ideas and the demands of chemical practice were in tension and eventual conflict. The surprising and revealing thing is how long Newtonian beliefs continued to dictate the research strategies of so many investigators. Even when, at the start of the nineteenth century, chemistry was equipped with alternative models and methods of demonstrated value, Newtonian aspirations lingered on.

6.2. Chemical Elements and Mechanical Philosophy

Late-Renaissance dissatisfaction with Aristotelianism resulted not only in ultimately triumphant endeavors to develop a mechanistic natural philosophy, but also in Paracelsian-inspired attempts to create a new Christian and chemical world-system.[1] In his 1661 *Sceptical Chymist,* such a mechanist as Robert Boyle was as much concerned to refute the arguments of the Paracelsians or "Spagyrists" as those of the Aristotelians or "Peripateticks." Indeed, reaction against the wilder excesses of the Paracelsian school provided an important psychological motivation for the mid-seventeenth-century drive to "beget a good understanding betwixt the [practical] chymists and the mechanical philosophers."[2]

This drive was so successful that in much of Europe the more

[1] See the detailed studies by Professor A. G. Debus, notably Debus (1965), *The English Paracelsians.*

[2] See Rattansi (1963), "Paracelsus and Puritan revolution." The quotation is from Hall (1965), *Boyle on Natural Philosophy,* 283.

purely chemical theorists had been driven from the center of the stage by 1700. Their defeat in this encounter, and the rapidly accruing prestige of the new Newtonian physics, go far toward explaining just why reductionist assumptions conditioned so much eighteenth-century chemistry. The belief that all chemical phenomena were best explained in physical terms itself mutated as physics changed from the corpuscularian or Cartesian and kinematic, to the Newtonian and dynamic. And so profound was this latter change that in 1709 John Freind could in all seriousness look back and declare that "Mr Boyl[e] . . . has not so much laid a new foundation of chymistry as he has thrown down the old."

Though Freind's own immoderate faith in attraction was by no means universally shared in the early eighteenth century, the move from mechanisms based on shape to ones based on interparticle forces did profoundly alter the physicalist interpretation of chemistry. A further change lay in Newton's adoption of constant inertial density as the standard of homogeneity for his universal matter, and his later insistence on the enduring nature of his primary particles. Even so, Newton, like Boyle before him, still held that these primary particles were both far below the observational level and of varied size and shape among themselves. Most important of all, Newton also took over and reinforced the prevailing mechanistic denial of fundamental importance to the chemical level of organization.

Boyle himself ably expressed this physicalist position by saying (in his *Excellency and Grounds of the Corpuscular or Mechanical Hypothesis*) that

> though chymical explications be sometimes the most obvious and ready, yet they are not the most fundamental and satisfactory: for, the chemical ingredient itself . . . must owe its nature and other qualities to the union of insensible particles in a convenient size, shape, motion or rest, and contexture; all which are but mechanical affections of convening corpuscles . . . [thus] these more confined hypotheses [the chemical

principles] may be subordinated to those more general and fertile principles [matter and motion], and there can be no ingredient assigned, that has a real existence in Nature, that may not be derived . . . from the universal matter, modified by its mechanical affections.[3]

Boyle went on to defend his position by using the analogy of many different buildings being constructed from the same bricks. Though Newton himself did not use it, this same analogy may be found sixty years later in Desaguliers's *Experimental Philosophy*—a point which well illustrates how indebted to its seventeenth-century background was Newtonian belief in a hierarchical internal structure to matter.[4]

While mechanistic approaches to chemistry denied fundamental importance to chemical elements, they did not go so far as to suggest that all chemicals were equal. In the *Origins of Forms and Qualities,* Boyle had explained how the *minima naturalia* could be compounded into "primitive concretions" which then formed "even grosser and more compounded corpuscles [which] may have . . . a permanent texture: for quicksilver, for instance, may be turned into a red powder . . . and disguised I know not how many other ways, and yet remain true and recoverable mercury." The idea of compounded particles of greater stability was also explored in the *Sceptical Chymist.* Similar ideas may be seen in Newton's Query 31 statement that "the smallest particles of matter may cohere by the strongest attractions, and compose bigger particles of weaker virtue . . . until the progression end in the biggest particles on which the operations in chymistry . . . depend."[5] And in the *De Natura Acidorum* he discussed at length the hierarchy of particles, and made it clear that gold could be transmuted *only if* its particles of the last, or next-to-the-last, composition could be broken down. In spite of these distinctions, the "mechanical philosophy" approach did not in itself offer a clear method for

[3] Quoted from Hall (1965), *Boyle on Natural Philosophy,* 200–203.
[4] *Ibid.,* 203–204, and Desaguliers (1734–44), I, 22–23.
[5] Hall (1965), *Boyle on Natural Philosophy,* 213–214, 211–213; Newton (1706), *Optice,* 337–338. Translation quoted from Newton (1718), *Opticks,* 370.

The Problem of the Elements

deciding which of all the known chemicals were fundamental, while it also carried the definite implication that further analysis *might* break anything down. The known chemical elements thus suffered a disturbingly insecure ontological status.

As the eighteenth century drew on, Newtonian natural philosophy increasingly supplied the, often unstated, theoretical background to chemical discussion. What this was to mean for ideas on the chemical elements was clearly foreshadowed in the 1704 volume of Harris' *Lexicon Technicum,* where under PRINCIPLE we read:

> In chymistry particularly, 'tis taken for first constituent and component particles of all bodies, out of which they are made, and into which they are by fire, as they say, resolvable again . . . the chymists did formerly pretend, that they could by their art resolve all natural bodies into these [the hypostatical principles] . . . But since this art hath been more commonly studied, and consequently much better known, it is found to be a falsity as Mr. Boyle excellently shows in his *Sceptical Chymist;* and Lemery hints in many places of his good *Course of Chymistry.* The modern chymists agree that there are five . . . *principles* . . . tho' these can never be drawn perfectly pure and unmix'd; nor have we any reason to believe they are the constituent principles of the bodies they are drawn from; and out of many bodies hardly ever a one of them can be drawn; and therefore they are not truly and properly the elements or constituent principles of natural bodies, nor indeed do we know any such.

Such skepticism was all very well for the natural philosopher. It was scarcely of comfort to the practicing chemist.

6.3. Homberg, Stahl, and Chemical Tradition

It may be well to begin this section by quoting the common mid-seventeenth-century definition of a chemical element, as it is found in a famous passage in Boyle's *Sceptical Chymist.* Boyle said:

> I now mean by elements . . . certain primitive and simple, or perfectly unmingled bodies; which not being made of any other bodies, or of one another, are the ingredients of which all those called perfectly mixt bodies are immediately compounded, and into which they are ultimately resolved: now whether there be any one such body to be constantly met with in all, and each, of those that are said to be elemented bodies, is the thing I now question.[6]

There are three distinct criteria for chemical elements, as defined in this passage. Such elements were: (a) "primitive and simple, or perfectly unmingled bodies," (b) "the ingredients of which all those called perfectly mixt bodies are immediately compounded, and into which they are ultimately resolved," and (c) "to be constantly met with in all, and each, of those that are said to be elemented bodies." Of course Boyle was concerned to argue that there were no bodies known that fulfilled all these criteria, and hence no bodies worthy of the name of chemical elements. What is of more interest here than the nature of Boyle's devastating critique is the continuing history of the term "chemical element."

The first criterion, that chemical elements were "simple," was quickly abandoned by those chemists who were willing to see their science reconciled with a mechanistic natural philosophy. Thus the Cartesian-inspired chemist Nicholas Lemery, in a discussion which appears to owe much to Boyle's critique, said "the word principle in chymistry must not be understood in too nice a sense: for the substances which are so called, are only principles in respect of us, and as we can advance no further in the division of bodies; but we well know that they may be still divided into abundance of other parts . . . wherefore such substances are to be understood by chymical principles, as are separated and divided, so far as we are capable of doing it by our weak imperfect powers."[7] Such statements were already

---

[6] Quoted from Hall (1965), *Boyle on Natural Philosophy*, 217–218.
[7] Lemery (1698), *Course of Chemistry*, 5–6.

The Problem of the Elements

current well before Newton's ideas, with their demand for internally structured and highly porous matter, were widely known or generally accepted. We can understand, therefore, why Newton's work was in due course thought to provide powerful additional support for a mechanistic approach to chemistry.

While the existence of an internal structure to the chemical elements was quickly conceded by many chemists, Boyle's second and third criteria were more sternly defended. Even so, acceptance of an internal structure, however reluctant, inevitably led to the conclusion that Boyle's third criterion must also be abandoned. The "all in all bodies" view, or the idea that *every* known substance contained *all* the chemical elements, was clearly unnecessary in a mechanistic world. For if the chemical elements were themselves only complex arrangements of primary particles, and if these elements were transmutable into one another, there was no obvious and *a priori* reason for expecting them all to occur in all bodies.

The "all in all bodies" view of the elements suffered a slow attenuation and decay as the eighteenth century progressed. This decay was caused at least as much by the loss of a firm base for the belief, as by practical difficulties in detecting all the elements in all bodies. Thus though Lemery defended this criterion, already in 1702 we find it abandoned by such an alert chemist as Wilhelm Homberg. Only Boyle's second condition was to survive the century unchanged. Here the continuing argument about whether the ingredients "into which they are ultimately resolved" really existed in the compound before analysis began was gradually answered in the affirmative, thanks to increasing ability not only to analyze but also to *resynthesize* compounds after analysis.

This ability was partly the result of a movement away from excessive reliance on the inevitably destructive technique of "analysis by fire." Such analysis ("destructive distillation" in more modern terms) could not but raise doubts about the preexistence of the products in the substance so harshly treated. A shift to less vigorous techniques was an important empirical step toward a *more easily reproducible* chemistry. Technical art was more important than alchemical faith when it came to such matters as precipitation by alkali and solution by acid as means of analysis. An excellent example of the power of such means may be seen in Joseph Black's 1756 paper on *Magnesia Alba*. As Black's classic investigations illustrate, analysis and synthesis techniques were also to gain enormously as the absorption and evolution of "airs" in reaction came to be systematically studied, rather than casually dismissed.

Before the middle of the century, and the arrival of gas chemistry, the definition of chemical elements was already reduced on occasion to "the ingredients of which all those called perfectly mixt bodies are immediately compounded, and into which they are ultimately resolved." The acceptance of such a working definition, and the desire to use analysis and synthesis techniques, may be seen for instance in Stahl's well-known "proof" that sulfuric acid is an element, from which the "compound" sulfur may be obtained by the addition of phlogiston. The whole new field of pneumatic chemistry in its turn proved well-suited to such an operational definition of chemical elements. Rephrased in a "negative-empirical" way, to stress the present and supposedly temporary limits of analytical technique, this definition became common in the later eighteenth century. It proved equally acceptable to Newtonians, empirics, and both supporters and opponents of Lavoisier's reforms. Indeed Lavoisier's famous experiments on mercury and air in a closed system served to show that the "new" acidifying principle was an element, by employing that same analysis and synthesis approach which had earlier served Stahl's sulfuric acid.[8]

To see more fully into the ways in which theoretically inclined practicing chemists adapted themselves to a mechanist-

---

[8] The technical details of the experiments undertaken by Stahl, Black, and Lavoisier may best be consulted in *Partington* (1961–64), II, 671–672, and III, 135–140 and 416 *et seq*. The phrase "negative-empirical" belongs to Dr. D. M. Knight. It is an apt way of characterizing the "what cannot be broken down" definition of chemical elements—a definition based on what cannot be achieved experimentally.

dominated world, we shall examine first Homberg, then Stahl. Wilhelm Homberg[9] was an influential French chemist of the early eighteenth century. A member of the Académie Royale des Sciences, and a friend of Nicholas Lemery, he traveled widely in Europe and at one stage worked in Boyle's laboratory. He made important quantitative studies on acid-base neutralizations, and his experiments with the large burning mirror that belonged to the Duke of Orleans had a marked effect on the development of eighteenth-century theories of light, fire, and combustion. Homberg was well known and aware of the new forces affecting the chemistry of his day. We may assume therefore that the series of "Essais de chimie" which he wrote was widely read. These essays in their turn represent a sustained attempt to reconcile the complexities of chemical experience with the demand for "clear and distinct ideas," so common among mechanist philosophers.

Homberg defined chemical elements solely in terms of Boyle's second criterion. He thus made what appears to be the earliest use of a purely "negative-empirical" definition of the sort that was to become commonplace later in the century. Homberg insisted that "the word principle . . . signifies only the simplest things into which a mixt is reduced by chemical analysis." This definition may be seen in use when, discussing salts, he says: "There are others we know to be mixtures, but it is not in our power to separate them. We take them for one of our chemical principles, because our analyses are not able to render them simpler, which is the nature of our principles." The abandonment of the "all in all" view appears in connection with his five chemical elements, for "different combinations of these five, or of some among them, makes the variety of all the substances it is in our power to examine." For instance, though all five elements occur in the mineral world, mercury is not an element of fossil salts, simple stones, or earths.[10]

[9] (1652–1715). See *Partington* (1961–64), III, 42–47.

[10] Homberg (1702), 33, 36, and 34–35. The "Essais," which appeared in the *Mémoires* of the French Académie, seem to have been intended as drafts for a never published textbook.

Atoms and Powers

Though in Homberg we already see the abandonment of the "simple substance" and the "all in all" positions, under the impact of mechanistic theories, we do not find any corresponding change in actual chemical conclusions. His negative-empirical definition leads him to five traditional elements—salt, sulfur, mercury, water, and earth—though under salt for instance, he does go on to admit three different salts (a fixed salt, a volatile or urinous salt, and an acid salt). His list shows how the "property carrying" nature of elements, though unstated, remained of great importance to the practical chemist. Thus Homberg was sure that mercury (the cause of luster, malleability, density, etc.) occurred in all the metals, but not in any vegetables or animals. Sulfur was likewise always present in inflammable bodies.

The question of quite how chemical specificity was to be explained was always a problem for mechanistic systems. It was also an unconscious reason for chemists to prefer a few elements which could be related to important chemical characteristics with a one-to-one correspondence. In each case the chemical principle was the carrier of the specific chemical property, the property itself being mechanically caused by the shape, size, and motion of the particles of the principle (and subsequently, for Newtonians, by the power of the particles' attractive force). Such marriages of mechanical and chemical explanations were obviously uneasy. It is thus not surprising that later French reductionist chemists were to differ substantially from their British counterparts on this very issue of chemical specificity.

The other great chemical problem which still confronted even a simple negative-empirical definition was that of deciding the actual members of the defined class. This may be illustrated by two examples. Homberg himself decided in 1705 that "la matière de la lumière" was actually sulfur (the true chemical element, that is). He adduced as confirming evidence the way in which light was always active and in motion, like sulfur, the active principle. More important, this light could be captured and fixed in bodies—as when regulus of antimony was calcined by a burning glass, or mercury heated in a glass vessel over a fire.[11] Clearly nothing but light (i.e., the sulfur principle) could enter the metal and account for its increase in weight or change in properties. Consider the further difficulties and confusions revealed by Peter Shaw's report on Homberg's researches. In his copious notes to Boerhaave's chemistry, Shaw tells how "even M. *Homberg,* who considers an element or principle in a stricter sense than many of the rest, *viz.* as a body which cannot by any analysis be reduced into simpler parts, treats mercury as an element: not but that he thinks it a compound; but because the method of decompounding it has not been yet discovered." The reason for Homberg's skepticism about the elemental status of mercury was that it "may be destroyed, which a body perfectly simple cannot," such destruction (volatilization) occurring when the mercury was subject to great heat! [12]

These instances make it plain how even the adoption of a mechanistically acceptable definition of a chemical element did little to help the immediate development of chemistry. The limited, positivistic, practical approach of Homberg was of some use. Clear thinking, careful experimenting, and specific factual information were admirable in themselves. But more needed was a philosophy that took account of mechanist demands, while insisting on and fully utilizing directly chemical insights. The formulation of such a philosophy, brought to full fruition in the new chemistry of the end of the century, was begun by Stahl.

Georg Ernst Stahl [13] is one of the most interesting, and most neglected, figures in the history of science. A pietist and a medical graduate of Jena, he belongs in an important Continental tradition of Protestant medical teachers and chemical experi-

---

[11] Homberg (1705), "Du Souphre Principe," 96. In modern terminology, Homberg's apparently quite reasonable argument becomes "that the reaction $2Hg + O_2 \longrightarrow 2HgO$ proves light is sulfur!"

[12] Boerhaave (1727), notes to pages 75 and 182.

[13] (1660–1734). See Metzger (1930), *Newton, Stahl, Boerhaave,* for the only serious twentieth-century study of Stahl's chemistry.

menters. Through Friedrich Hoffmann, his professorial colleague at Halle, and later through Caspar Neumann, Stahl was undoubtedly aware of the direction of English mechanistic thought in both its Boylean and its Newtonian forms. Such ideas had little appeal for him. The reason was that "mechanical philosophy, though it vaunts itself as capable of explaining everything most clearly, has applied itself rather presumptuously to the consideration of chemico-physical matters . . . it scratches the shell and surface of things and leaves the kernel untouched." [14] Instead, resort must be had to more directly chemical traditions if chemical insight was to be obtained.

Drawing on such a varied chemical background as that provided by the works of his teacher, the alchemist G. W. Wedel, the Helmontian textbook of Jacob Barner, and of course the writings of J. J. Becher, Stahl sought to remodel chemical theory in a way that was at once rational, empirical, and true to chemical experience rather than physical theory. The subtlety and variety of his work has since been hidden by early twentieth-century commentators and their obsession with his "erroneous" phlogiston theory.[15] His undeniable influence on the course of chemistry—especially German chemistry—is therefore still largely unexplored. An additional reason for this neglect is the rarity, prolixity, and obscurity of much of his voluminous output. An over-all assessment of Stahl's work cannot concern us here. But his theory of matter is of particular interest for the way in which it recognized the validity of a unity-of-matter, internal-structure approach, yet legislated it out of the field of chemistry.

Stahl's work thus admitted of Newtonian and positivist as

[14] Quoted from *Partington* (1961–64), II, 665.

[15] An attitude best revealed in Partington's contrast of "the path of true discovery opened out by Boyle, Hooke and Mayow" with "the jungle of the Theory of Phlogiston": Partington (1937), *Short History*, 84. Revealing a similar bias, Professor Butterfield writes disparagingly of "[those] modern historical writers [who] have tended to try to be kind to the phlogiston theory, apparently on the view that it is the historian's function to be charitable, and that the sympathy due to human beings can properly be extended to inanimate things": Butterfield (1949), *Origins of Modern Science*, 180.

The Problem of the Elements

well as more directly chemical interpretations. More important still for chemistry, Stahl definitely accepted "property-bearing" elements. This acceptance, probably owing much to his Paracelsian predecessors, was a chemical necessity at a time when acidity, alkalinity, combustibility, and metallicity were all becoming more sharply defined and more important in chemical thought. Practicing chemists had a real need for explanations of these phenomena. Naturally explanations in terms of definite material substances were favored. To say that the cause of color lay in the phlogistic principle and to relate particular colors of particular substances to their relative combustibility (also caused by this same phlogistic principle) was to offer the prospect of *useful* explanations of a whole range of chemical phenomena. The rival Newtonian scheme of explanation by differing attractive forces, though always of great interest to the reductionist-minded and theoretically inclined, did not possess the same immediate appeal to practical chemists.

The central importance of matter-theory to Stahl's approach is apparent from his exposition of the "General theory of chemistry." This begins with eighteen pages on "the structure of simple, mix'd, compound and aggregate matter." In his system, the principles or simple bodies which "do not consist of physical parts" (i.e., are not divisible) are fundamental. These principles are "the first material causes of mixts; and the compounded, according to the difference of their mixture, are either mix'd, compound or aggregate: mix'd if composed merely of principles; compounds, if form'd of mixts into any determinable single thing; and aggregate, when several such things form any other entire parcel of matter, whatsoever it be."[16] In other words, we have a structure of increasing complexity as:

Principles ⟶	Mixts ⟶	Compounds ⟶	Aggregates
[Simple bodies]	[2 or more principles]	[2 or more mixts]	[2 or more compounds]

[16] Stahl (1730), *Principles of Universal Chemistry*, 2–3.

Atoms and Powers

In agreement with both physicalist and more directly chemical ideas, Stahl defined principles as "a priori, that in mix'd matter, which first existed [the Newtonian view]; and a posteriori, that into which it is at last resolved [the chemical view]." However he went on to make it quite plain that "a difference, at present, prevails between the physical and chemical principles of mix'd bodies. Those are called physical principles whereof a mixt is really composed; but they are not hitherto settled . . . And those are usually term'd chemical principles, into which all bodies are found reducible by the chemical operations hitherto known." This statement of the difference between physical and chemical principles was to prove of great importance. It was to be endlessly repeated by later writers on chemistry.

Because Stahl accepted the Boylean definition that *all* chemical elements should be present in *all* bodies, he, like Boyle, denied the validity of both the Aristotelian and the Paracelsian elements. Boyle had wished his own skepticism to promote an alternative corpuscular philosophy. Stahl was more firmly within a chemical tradition. He therefore went on to argue that "all the darkness and disputes about principles arise from a neglect of that real distinction between original and secondary mixts, or mixts consisting of principles and bodies compounded of mixts." Stahl's particular conviction was that "by justly distinguishing between mixts and compounds, without directly undertaking to exhibit the first principles of the latter, we may easily settle this affair." [17] The point was that in Nature one always dealt with compounds. It was therefore to be expected that the substances obtained upon their analysis (mixts) should differ. The existence of two separate oils, obtained by distilling two different twigs, was no longer a scandal—as it had been for Boyle—but merely indicated that oils were "mixts," potentially resoluble into their principles.

Stahl's work well represents the ambiguities and uncertainties facing the more theoretically inclined chemists of the

[17] *Ibid.*, 4 and 5.

The Problem of the Elements

period. Aware of, and accommodating, the hierarchical structure of matter that mechanistic philosophies demanded, Stahl yet wished to preserve definite "chemical elements." Clearly only the last products of analysis could qualify, *yet* Stahl's elements turn out to be water and three types of earth. As natural philosopher, Stahl himself was unhappy about this. He admitted that "this hypothesis also, of four principles, is not strictly true," though "we may . . . very well allow it as the most probable, and till time shall make farther discoveries, retain it for the better explanation of chemical operations and phaenomena." [18]

As this last phrase shows, it was the chemical utility of these principles which appealed to Stahl. Vitreous earth, sulfureous earth, mercurial earth and water between them offered definite material bases for every fundamental and recurring chemical quality. To equate each and every case of salt-like character with the presence of a specific vitreous earth, each observation of inflammability with the sulfureous, was to offer a considerable and persuasive increase in chemical explanatory power. This explanatory power was to ensure a wide reception for similar ideas among later chemists. In theory, an analysis such as Stahl's opened the way to reconciliation of physicalist and chemical belief. His admission of an internal structure to matter, and his clear differentiation of physical from chemical principles, should have removed the tensions between a wish to pursue chemical experience and a desire to explain that experience in physical terms. In practice it was quite otherwise.

In Britain the Newtonian chemical tradition that flowed from Freind and Hales, through Cullen and Black, to Priestley and Kirwan, engendered a suspicion of attempts to give special

[18] *Ibid.,* 10. Two further related issues that cannot be explored here, but that will be touched on from time to time in this chapter, are whether *in fact* the known chemical elements can be transmuted into one another, and whether the chemical properties of a compound are the mean of, or wholly distinct from, those of its constituent elements. In denying transmutation and accepting property-bearing elements, Lavoisier was to reveal debts to the Stahlian tradition. That similar debts were not owed by British Newtonians such as Cullen will become apparent from what follows.

status, and property-bearing qualities, to any particular group of elements. While admissible in the short term, such efforts could only militate against the true Newtonian quest for fundamental physicomathematical explanation, and obscure how "there is no reason to suppose any real indestructible principle has yet been discovered." In France the position was reversed. Lip service might be paid to the idea that "these things we suppose to be simple may soon be found quite otherwise." In practice, increasing sympathy was evident for Stahl's belief that mechanistic interpretations scratched the surface of things, while leaving the kernel untouched. The ablest chemical theorists, from Venel and Macquer to Lavoisier and Fourcroy, were all deeply influenced by the need to admit and use definite elements or principles with property-bearing characteristics related to the realm of laboratory experience.

6.4. Particle Size and the "Nut-Shell" Theory
While an uneasy truce between physicalist assumptions and chemical necessities might be achieved in such a scheme as Stahl's, the physicalist background was itself by no means fixed and immutable. The difficulties of any exploration of the internal structure of matter became ever more apparent with each fresh piece of evidence on the incredible minuteness of particles of chemical compounds, let alone physical atoms. Such knowledge of the problems was reinforced and strengthened by the continued development and widespread diffusion of the "nut-shell" theory.

A new interest in the actual sizes of the smallest known particles of matter was a natural consequence of seventeenth-century attempts to develop a satisfactory mechanical philosophy. Robert Boyle was particularly active in this field. He produced many arguments for the almost unbelievably small size of ultimate constituent particles. His experiments on the coloring power of solutions, and weight-loss versus penetrating ability for powerful odorizers, became standard fare for later commenta-

The Problem of the Elements

tors. These commentators usually coupled Boyle's earlier work with Edmond Halley's 1693 "Demonstration of the exceeding minuteness of the atoms or constituent particles of gold."[19]

As befitted the publisher of the *Principia* and friend of Newton, Halley had been concerned with the question of "what are the constituent particles of matter, and how there comes to be so great a diversity in the weight of bodies to all appearances equally solid and dense, such as are gold and glass, (whose specifick gravities are nearly as 7 to 1)." Naturally Halley accepted that gravitational weight was proportional to quantity of matter. His paper thus offers a fascinating preview of some of the problems relating to the theory of matter, that Newton himself was publicly to explore at a later date. From his Newtonian position, Halley argued that "at least six parts of seven in the bulk of glass, must be pore or vacuity." And "this some favourers of the atomical philosophy have endeavoured to solve, by supposing the primary or constituent atoms of gold to be much larger than those of other bodies, and consequently the pores [between them] fewer."

This position was crude in the extreme, compared with Newton's later analysis of variation in *internal* structure. Nonetheless, it was sufficient for Halley's immediate purpose—showing the amazing minuteness of even these supposedly "much larger" atoms. Halley's "experiment" was simplicity itself. By studying the ability of gold to gild fine-drawn silver wire, he was able to show that, even were the gilding only one atom thick, it must still be the case that "the cube of the hundredth part of an inch [of gold] would contain above 2433000000 (or the cube of 1345) of such atoms."[20] Clearly, gold, though possibly the largest of that class of particles on which "the operations of chemistry" depended, was well below the observational level.

[19] See e.g., Desaguliers (1734–44), *Experimental Philosophy*, I, 27–30; Martin (1735), *Philosophical Grammar*, 42–44. Arguments about the small size of gold particles, and statements about condensing the earth until "smaller than a walnut" may be traced back at least to Galileo's *Dialogues Concerning Two New Sciences*.

[20] Halley (1693), 540–542.

Atoms and Powers

In due course the *Opticks* was to supply not only additional evidence on the minuteness of the constituent particles of matter, but also powerful arguments about their extreme porosity. As was pointed out above in sections 2.3 and 3.4.1., Newton's whole theory of the transmission of light through bodies assumed such porosity, together with the possession of *internal* structures. When we consider Newton's demonstration that "particles of the last composition" must be highly porous, alongside Halley's proof that even in the case of gold such particles are incredibly minute, it is not surprising that the still more remote "ultimate primary particles or atoms from which matter is composed" came to seem increasingly "metaphysical" as the eighteenth century progressed. Indeed the implications for chemistry of Newtonian natural philosophy were such that it is surprising only that the two did not part company far sooner.

One minor illustration of later belief in the remoteness of fundamental particles may be seen in a popular text by Benjamin Martin. Accepting the particulate nature of light, Martin argued that "in a second of a minute, there flies out of a burning candle, the following number of the particles of light, 41866000000000000000000000000000000000000000 which is 10000000000 or ten millions of millions a bigger number than 100000000000000000000000000000000, [which is] the number of the grains of sand computed to be contained in the whole earth." Since of course the weight-loss of the candle in any "second of a minute" was negligible, the implication about the quantity of matter in one light particle was plain. In similar fashion, Martin was able to harp on the Newtonian "nut-shell" theme, and point out how "the matter of transparent bodies must be very little in proportion to their pores ... so little, that a celebrated philosopher questioned whether the quantity of matter in glass were more in proportion to its porosity, than one grain of sand to the bulk of the whole earth." [21]

---

[21] Martin (1735), *Philosophical Grammar,* 60 and 80. Martin's printer seems to have been defeated by such an array of noughts, losing three from the middle number (i.e., the correct statement would assert that 10^{45} is 10^{13} larger than 10^{32}).

The Problem of the Elements

The most serious physicalist implications for thought on the chemical elements were those carried by the "nut-shell" theory. In the 1720's and 1730's, this view of matter seems to have shed its controversial origins and overtones and become widely accepted as a routine part of Newtonian natural philosophy. Evidence may be found not only in Dutch and British texts but also in the writings of Voltaire. His 1733 *Letters Concerning the English Nation* do not only contain that delightful passage we have already referred to, about how "a Frenchman who arrives in London, will find philosophy, like every thing else, very much chang'd there. He had left the world a plenum, and now he finds it a vacuum." Voltaire was also at pains to point out how by "examining the vast porosity of bodies, every particle having its pores, and every particle of those particles having its own; he [Newton] shows we are not certain that there is a cubic inch of solid matter in the universe, so far are we from conceiving what matter is." Similar passages may be found in Voltaire's later popular primer, the *Elements of Sir Isaac Newton's Philosophy*.[22]

Such awareness of the "nut-shell" theory, and of experiments on the amazing subtlety of the smallest constituent particles of any known chemical, could not but affect later chemical thought about the nature and status of the chemical elements. Earlier debate about these elements had been widely influenced by rival Aristotelian, Paracelsian, and corpuscularian cosmological schemes. In a similar fashion this later discussion was greatly influenced by the Newtonian *Weltanschauung*. However the possible modes of influence were more than one, as comparison of Dutch and British with French thought will show.

6.5. Elements and Atoms in Dutch and British Chemistry

One way in which Newtonian natural philosophy affected thought about the chemical elements may be seen very clearly in a question and answer passage in Benjamin Martin's *Philosophical Grammar*. The passage runs:

[22] Voltaire (1733b), 147, and (1738b), 110–111 and 162.

A. Is the original matter of all bodies the same?

B. Yes; what the philosophers call the primary constituent particles, atoms, or corpuscles of matter, which constitute bodies of infinite kinds, are yet the same, or of the same nature, among themselves, in all those bodies.

A. And what, is not this the same thing as to say, that fire and water, a flint and down, gold and dung, are the same things with respect to the matter itself of which they consist?

B. Yes, they are so; and what then, my friend?

The point was rammed home in a subsequent passage which reads:

A. What do you call the elements of natural bodies?

B. Those pure and simple substances of which all gross and mixed bodies are said to consist; and into which they may ultimately be resolv'd, or reduc'd.

A. How many are those elements reckoned to be?

B. The ancients counted seven . . . Some of the modern chemical philosophers reckon five . . . Whereas in reality, there are no other elements of natural bodies than the primogenial particles of matter, or substance of which they consist universally." [23]

Of course all of this might have been said by a Cartesian or a "corpuscularian." But the problems Newtonian natural philosophy posed for chemistry did not end here. It was one thing to insist that the true elements of bodies are "the primogenial particles of matter . . . of which they consist universally." It was quite another to say anything useful about them. As should by now be clear, the most salient facts about the "primogenial particles" were their almost unbelievable minuteness, and their inaccessibility to observation. This inaccessibility did not prevent highly competent and respected mid-century Newtonians from

---

[23] Martin (1735), 32 and 36.

The Problem of the Elements

discussing at some length how known chemicals were made up from such particles.

Peter Musschenbroek's *Elements of Natural Philosophy* explained in the orthodox way the process by which ultimate solids combine with space to give particles of the first order, then by a repeat process particles of the second order, and so on till "at length is composed a large and tractable body." (Even a diagram of the internal structure of matter was provided—see Fig. VIII.) The chemical implications were carefully explored:

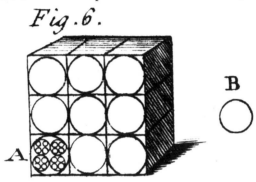

VIII. Musschenbroek on internal structure. From Musschenbroek (1744), I, plate I.

"Particles of the first order may be very different [from one another] in bigness, figure, porosity, density, gravity, coherence," with the consequence that "particles of the second order may be still more different . . . so that an infinite variety of bodies, differing as to figure, magnitude, gravity, coherence, and density may arise from such orders."

Granted such a variety of components, Musschenbroek found it easy to classify homogeneous bodies as "composed only of [identical] particles of one order." Heterogeneous bodies were ones in which "the first, second, third, &c. orders concur together, as also orders different from one another, as to figure, magnitude or density." Such a scheme agreed well with the way "experience informs us, that almost all the larger bodies are very heterogeneous and mixtures as it were of very different things. For almost all chemists in general instruct us, that metals are compounded of salt, sulphur, and mercury. And they likewise shew, that salt and sulphur are still heterogeneous compounds." [24] This analysis might be of no great help to the practical chemist in the laboratory. Still, it did convincingly relate the diversity of chemical experience to the increasingly widespread assumptions of the Newtonian natural philosophy. Of course it did far more than that. It unavoidably implied the unsatisfactory nature of chemical elements that were themselves highly complex. And it highlighted the superficiality of a chemistry that could say nothing about the internal structures on which these elements depended.

Such an acute and sophisticated chemist as Herman Boerhaave was well aware of the difficulties that lay ahead. Boerhaave, whose work we have already discussed in section 4.3.3, was the first and perhaps the most important of the continuing school of Dutch Newtonians. His *Elementa Chemiae* provides an excellent example of the impact of Newton's work on chemical thought, as well as being a classic in its own right. Boerhaave was convinced that chemical analysis did not yield "the exceeding fine elements" of which bodies were ultimately composed. He argued that "the parts into which the greatest masters pretend to have resolved compound bodies, are not themselves of a simple nature, but unstable and capable of farther division." The one possible exception was that "fire perhaps, and that only whilst it passes through gold, or the like substances, may give us its elements perfectly pure." [25] Yet even this belief was not without difficulty, for to Boerhaave fire and light were synonymous terms.

Accepting and arguing for the Newtonian doctrine of the porosity and hierarchical internal structure to matter, he thought that "the ultimate elements of fire, appear to be of such a nature, as to be the most solid, perhaps, of all bodies." Indeed fire, "according to this doctrine, will be totally corporeal, im-

[24] Musschenbroek (1744), *Natural Philosophy*, I, 31, 33, and 34. See also section 4.3.2. above.
[25] Boerhaave (1735), I, 46–47.

The Problem of the Elements

mutable, [and] incapable of having its figure changed." Even so, it had to be admitted that "an absolute simplicity of fire is repugnant to the doctrine of the great *Newton,* whose uncommon genius seems to have penetrated almost beyond the limits of human understanding. For this noble author, by an artificial separation of one ray of fire, has divided it into seven different ones."

If Newton could do this, "who knows what additions may be made hereafter to the Newtonian doctrine? . . . If, hereafter, now, this science should be more subtly cultivated, and dioptical instruments should be carried to greater perfection, who will pretend to assert, that even in these simple Newtonian rays, some penetrating genius may not be able to discover a still further composition?" In fact Newton himself had shown a further property of the rays. This was the diversity of their opposite sides (i.e., the phenomenon of double refraction). Boerhaave was driven to conclude that

> in this so vastly simple being, therefore, we see there still remains this manifold variety: What diversity, therefore, have we reason to expect in compounds? In the smallest bodies, we every where observe a resemblance of the greater. Had this discovery, which was reserved for the great Newton alone, still lain in obscurity, I don't doubt that we should all even at this time, have firmly believed that in a ray of light there was somewhat ultimately small, and infinitely simple: But convinced by his doctrine, we are now obliged to confess, that, tho' fire is of all known bodies the most simple, yet even in this, there is found to be a various multiplicity.[26]

It would be difficult to find a passage that better illustrates the impact of Newton's optical discoveries on chemical thought, quite apart from the influence of the "nut-shell" view of matter with which these discoveries were so intimately associated.

---

[26] Boerhaave (1735), I, 230, 232–233. See also his comments as quoted in section 4.3.3. above.

Atoms and Powers

The net effect of Newton's work on white light, his views about the nature of transparent substances, and the "nut-shell" theory, was so suggest to all Newtonian-inspired chemists that a deep gulf lay between chemical elements and ultimate physical atoms. Chemists might wish to learn of the latter, but they must of necessity be content with knowledge of the former. Stahl's formulation allowed the problem to be shelved, in favor of concentration on known chemicals. In contrast, Boerhaave's perceptive comments are indicative of what was to prove a continuing Dutch and British concern, not only with the immediate realm of chemical experience, but also with the underlying physical reality.

The writings of Peter Shaw indicate how the British practical and manufacturing chemist of the mid-century might acknowledge Newtonian formulations, and yet ignore them in practice. Shaw was well aware that "the more intelligent among the modern chemists do not understand by *principles* the original particles of matter, whereof all bodies are by the mathematical and mechanical philosophers supposed to consist." The reason was that "these particles remain indiscernable to the sense . . . nor have their figures and original differences been determined by a just induction." In a voice not untouched with irony, Shaw could say that "genuine chemistry" preferred to leave "to other philosophers the sublimer disquisitions of primary corpuscles, or atoms," and instead of such "metaphysical speculations" contented itself "with grosser principles, which are evident to the sense, and known to produce effects." This formulation betrays not only his knowledge of Boyle, Newton, and Stahl but also of the pressures real to a chemical manufacturer if remote from a matter theorist.[27]

While Shaw was representative of a new and growing group within eighteenth-century Britain, the main debate still lay with Newtonian natural philosophers, theological inquirers, and chemists within university medical faculties. How much

[27] Shaw (1734), *Chemical Lectures,* 146. On Shaw's technological interests, see the study by Gibbs (1951b), "Peter Shaw and the revival of chemistry."

The Problem of the Elements

this last group—still the most important source of advanced chemical instruction—inclined toward Newtonian explanations may be seen from William Cullen's unpublished but widely influential lectures. To quote from a set of student's notes of 1762:

> Elements are physical or chemical, the former are the real elements of bodies or as they are often called atoms, but these physical elements are rather imagined than actually known [though "Sir Isaac Newton has furnished us with a good argument for the existence of these elements"] ... the strict, precise meaning of element is, that which no human art can divide; & those we call chemical elements but physical elements are those beyond which no power in our system can go ... All the bodies obvious to our senses, are compounded through several degrees ... art never attains the ultimate degree of division, but rests in some of the intermediate stages, which we may consider as the elements, only of a higher composition; hence, however, chemical elements.[28]

This sort of Newtonian analysis carried definite implications on such important matters as transmutation, and the acceptability of property-bearing chemical principles. In his lectures, Cullen went on to explore these implications. His arguments displayed clarity of thought, lucidity of expression, and catholicity of knowledge in a way that could not fail to impress his listeners. Ranging easily from Shaw, Stahl, and Boerhaave to

---

[28] Manchester University Library, Special Collections, Ms. no. C H C 121 1–4, I, pp. 13–15. (Cited hereafter as Cullen [1762].) These student notes, which are carefully copied in a clear hand, became the property of William Henry, who himself began his student career at Edinburgh in 1795. See Smith (1856), *Memoir of Dalton*, 144. Cf. Wellcome Historical Medical Library, Ms. no. 4674–4675, p. 1. An extensive critical study of the still-surviving notes from Cullen's lectures is much to be desired: a preliminary investigation was made in Wightman (1955–56), "Cullen and chemistry." As will be apparent from the present excerpts, Cullen was a man of wide reading and considerable intellectual power. His lectures thus throw a good deal of light on Scottish chemistry in its great creative phase. Henry's purchase of a thirty-three-year-old set of the lectures is testimony to their enduring worth and influence.

Atoms and Powers

Macquer and Venel, Cullen argued consistently and powerfully for a physicalist and Newtonian interpretation of chemical phenomena. His popular and influential university courses, together with those of his most famous pupil, Joseph Black, go far toward explaining how British chemical thought was so strongly set in a physicalist mold in the later eighteenth century.

Cullen was quite clear that "the end, and purpose of chemistry is the knowledge of the ultimate causes of qualities existing in bodies." In keeping with his Newtonian position, he further maintained that "all the operations of Nature and Art depend upon attraction and repulsion." He was even prepared to allow—though not to insist on—the idea that the repulsive-fluid matter of heat was "the only active principle in Nature," and to endorse Bryan Robinson's systematics.[29] Against this background, his discussion of chemical reaction and his answer to the question of "to what must we refer these changes of qualities [observed in reaction]," are entirely predictable—but none the less important for that. As a committed reductionist, Cullen insisted that "all the changes in the qualities of bodies are produced by combination or separation," that "the qualities of the constituent parts cannot appear in the mixt," and that "this general proposition . . . is the foundation of the theory of chemistry, which is the theory of the qualities of bodies."[30]

In order to underline his beliefs, and to remove all room for doubt and confusion, Cullen discussed the use of "substantial forms or hypostatical principles," i.e., certain qualities inherent in particular elements which are carried along with "[th]em into every degree of composition." His condemnation of the theory, and of "some men of knowledge and ingenuity [who] even at this very time give in to the Peripatetick system," was vigorous. Also typical was his attack on those who did such things as ascribe the cause of color to the coloring agent. Even worse were those "chemists [who] have referred . . . colour to the phlogiston. Now this is confounding all knowledge and philosophy, for without doubt colours arise from the size, figure, and

---

[29] Cullen (1762), I, 2 and 44–45. See also section 5.3.1. above.
[30] *Ibid.*, 30–31.

The Problem of the Elements

texture of the parts of matter."[31] We can imagine what must have been his reaction to Lavoisier's ascription of acidity to the presence of oxygen, the acidifying principle!

Though impatient with all but physical explanations, Cullen was yet chemist enough to admit that "in practice, it is always of consequence to know what substance will communicate the qualities we want, as in the communication of colours in dy[e]ing."[32] In such an admission as this, we once more see industrial pressure toward a chemistry that related to experience. However intellectually satisfying the belief that chemical elements were remote from physical atoms, and that "all the operations of Nature and Art depend upon attraction and repulsion," Newtonian theory had embarrassingly little of definite value to offer the practical chemist.

Though aware of the chemical utility of principles, Cullen was unremitting in his hostility toward explanations that rested with them, or with analogous chemical elements. To admit that there were some chemical elements which "no human art can divide" was not to give these elements any privileged status or security for

> the chemical elements are really *transmutable;* water may be transmuted into earth, mercury into water. If there are permanent and fixed qualities annexed to elements, these elements should carry these qualities into every mixt of which they make a part. But this is contrary to universal observation . . . It appears then that we know of no physical element, nor any chemical principle; nor are we acquainted with any body which hath fixed and permanent qualities. The doctrine of elements therefore, is not only useless, but prejudicial.[33]

---

[31] *Ibid.,* 31–33.

[32] *Ibid.,* 32. For Cullen's own extensive indusrial commitments, see Clow and Clow (1952), *Chemical Revolution, passim.*

[33] Cullen (1762), I, 33–34. As every eighteenth-century figure deeply involved in chemical practice, in contrast to those with primarily a theoretical interest, Cullen did of course have to settle for and use a determinate number of known simple chemicals. What is significant is his deliberate and theoretically motivated refusal to accord those chemicals a unique status, or to manifest particular interest in their *chemical* properties.

Atoms and Powers

Such an attitude was undeniably in accord with the strict canons of Newtonian matter-theory, but what must strike a later reader is its total lack of *chemical utility*. The statement is also of considerable interest for the light it throws on Lavoisier's work later in the same decade. His deep desire to prove the chemical elements not transmutable was anything but part of a physicalist "Postponed Scientific Revolution in Chemistry."

The strength of Newtonianism in Britain, its continuing relationship with theology (in both advanced and conservative forms), and its underpinning of sophisticated chemical theory, all help to explain the welcome accorded to Boscovich's work by British thinkers later in the century. Boscovich himself visited Britain in 1760. His ideas seem to have found a particular welcome in the Scottish universities, concerned as they were to counter the atheism and skepticism of Hume's philosophy. To modern eyes, Boscovich's work may sit ill with "common-sense" philosophy. To contemporaries, the use made of it by that "honorary" Edinburgh man, Joseph Priestley, was far more scandalous. However both cases agree in revealing how closely interwoven with philosophical, theological, and even political ideas was the apparently scientific question of the number and nature of the chemical elements.

It was Boscovich's *Theoria Philosophiae Naturalis* that finally took the logical step of emptying the Newtonian "nut-shell" of whatever "solid matter" it still contained. The step was taken with a lively concern for the theological implications of the resulting world-picture, as both the main text of the *Theoria* and its appendix reveal.[34] Boscovich was at some pains to preserve a matter-spirit dualism in his new theory. This should not surprise us. He was after all a highly placed Jesuit and a devoted

---

[34] As so often with eighteenth-century texts, it is possible to read these parts as mere decoration. That this would be a serious mistake is apparent from Boscovich's own later angry insistence on how "I had declared in my work, with all the clarity and precision possible, my . . . beliefs in speaking of the great difference which exists between matter and spirit, and in writing all that I have in my appendix on the soul and God": see Schofield (1967), *Scientific Autobiography*, 169.

servant of his church. No more should it surprise us that his treatise met with little response in the hostile climate that France then provided, or that natural philosophers within the theologically orientated British tradition greeted his ideas with enthusiasm.

Most enthusiastic of all was Joseph Priestley.[35] His 1772 *History of Vision* referred at length to Boscovich as "one of the first rate mathematicians and philosophers of the present age . . . who has particularly distinguished himself by a new and general theory of the laws of Nature." On his visit to Paris two years later, Priestley actually met Boscovich. As subsequent events were to show, communication between the English Dissenter and the Continental Jesuit was somewhat less than total. Nonetheless Priestley's 1777 *Disquisitions Relating to Matter and Spirit* were to announce his conversion to a Boscovichean view.

Priestley was at pains to point out how "it was a considerable time before I could bring myself really to receive a doctrine so new, though so strictly philosophical." In this, he was referring to the startling theological conclusions to be drawn from the new analysis, rather than to its scientific component. For as the logic-loving Priestley was to point out in defense of his own abandonment of solid matter, he was but completing the process initiated by Newton and Keill. Priestley argued:

> The principles of the Newtonian philosophy were no sooner known, than it was seen how few in comparison, of the phenomena of Nature were owing to solid matter, and how much to powers . . . It has been asserted . . . [that] all the solid matter in the solar system might be contained within a nut-shell . . . Now when solidity had apparently so very little to do in the system, it is really a wonder that it did not occur to philosophers sooner . . . that there might be no such thing in Nature.[36]

[35] (1733–1804). See Schofield (1966), "Priestley—philosopher," (1967) *Scientific Autobiography*, and Gibbs (1965), *Priestley*.
[36] Priestley (1772), I, 308, 383–394, etc., and (1777), x and 17.

Atoms and Powers

As even a cursory perusal of the *Disquisitions* will reveal, it was primarily for theological reasons that Priestley welcomed this extreme development of the Newtonian position. Unlike his mentor Boscovich, the theologically advanced and rationalistically inclined Priestley wished the disappearance of solid matter to signal an end to matter-spirit dualism. Predictably, Boscovich was bitterly offended by the use made of his ideas. He voiced the sentiments of orthodox churchmen everywhere when he rapidly objected to Priestley's "pure and unconcealed materialism" and to being made a "party to a doctrine that I detest and abhor as impiety in religion and senseless to sound philosophy."[37] To call Priestley's matterless (and therefore dualism-denying) universe materialistic was to reveal just how the argument had shifted from eighty years before. Newton's vision of forces as God's guarantee, and matter as His enemy, was strangely inverted by the work of Priestley. Stranger still was to be the reaction of British Evangelical Protestants.[38] But here and now the significant point is that both Boscovich and Priestley were agreed on the relevance of theories of matter to theology. Also significant is their agreement as to the hierarchical internal structure of matter, the complex nature of the known chemical elements, and the almost total vacuity of the universe.

The chemical discussion of the *Theoria* demands brief citation. Boscovich first showed how three or more of his dimensionless points might interact to give stable "particles of the first order." From these, "particles of the second order" resulted "and so on; until at last we reach those . . . variable particles, which are concerned in chemical operations . . . with regard to which we get the very thing set forth by Newton, in his last question in Optics, with respect to his primary elemental particles, that form other particles of different orders." Boscovich then went on to provide a sophisticated discussion, not only of the mechanism of light transmission and chemical reaction, but also of the internal structure of the known chemical elements. One

---

[37] Schofield (1967), *Scientific Autobiography*, 169.
[38] See section 8.3. below.

of his final, summarizing paragraphs pointed out how "if we could inspect the innermost constitution of particles and their structure . . . we should find some classes of particles to be so tenacious of their form that in all changes they would never be broken down . . . It would then be possible to divide with far greater certainty bodies into their species, and to distinguish certain elements which could be taken as the simple elements, unalterable by any force in Nature." [39]

Just such a view, coupled with more recent experimental successes, led to Priestley's well-known 1793 declaration of Newtonian chemical faith. His optimistic view that "our powers of investigation seem . . . to increase without limits" was also typical of the period. It found echo thirteen years later in the sentiments of another British chemist who was also to prove susceptible to the ideas of Boscovich. Humphry Davy, in proud possession of a great voltaic pile, was quite convinced that "the powers of our artificial instruments seem capable of indefinite increase." And as for Priestley, so for Davy, "there is no reason to suppose that any real *indestructible principle* has yet been discovered." [40]

Despite the use Priestley made of them, Boscovich's ideas found considerable favor with British moral as well as natural philosophers. This was particularly true in the Scottish school of Reid, Beattie, and Stewart.[41] Among the natural philosophers, we may note an early and favorable reference to Boscovich by Black's student William Cleghorn, and Dugald Stewart's later testimony that John Robison (Black's successor at Glasgow, and Edinburgh professor of natural philosophy from 1773)

---

[39] Boscovich (1763), pars. 239 and 531.

[40] Priestley (1793), *Generation of Air*, 38–39. The passage runs in part: "The advances we are continually making in the analysis of natural substances into the elements of which they consist, bring us but one step nearer . . . the knowledge of the internal arrangements of the elementary parts of natural substances . . . Nature exhibits an infinitely large field, but our powers of investigation seem also to increase without limits." Davy's ideas are most conveniently consulted in Davy (1839–40), *Works*, V, 54, and IV, 132.

[41] The author is indebted to Professor R. Olson for letting him see an unpublished paper which explores in detail "The reception of Boscovich's ideas in Scotland."

"had a strong and avowed leaning to the physical theory of Boscovich."[42] Thomas Thomson was a pupil of Black and Robison. We can therefore appreciate why he was so knowledgeable on and partial to Boscovich in the bestselling chemical textbook he launched on the world in 1802.

How much the thinking of British theoretical chemists ran on Newtonian and Boscovichean lines at the turn of the century may be seen in Thomson's statement that

> as the term simple substance in chemistry means nothing more than a body whose component parts are unknown . . . when the science reaches the highest point of perfection . . . the number of simple substances will probably be much smaller than at present. Indeed, it has been the opinion of many distinguished philosophers in all ages, that there is only one kind of matter . . . This opinion was adopted by Newton; and Boscovich has built upon it an exceedingly ingenious and instructive theory.[43]

However ingenious and instructive the theory, it was not to provide the means to chemical advance.

6.6. The French Situation

If a thoroughgoing Newtonianism was the hallmark of chemical thought in eighteenth-century Britain, the situation in France was by no means so simple. Indicative of the difference in mood is the way that Keill's and Freind's work appeared in French translation in 1723 as but part of the *Nouveau Cours de Chymie, Suivant les Principes de Newton & de Stahl*. We have already discussed Newton's influence. But it is impossible to understand the course of French chemistry without also considering Stahl.

---

[42] See Heathcote and McKie (1958), "Cleghorn's *De Igne*," 11; Hamilton (1854), *Stewart's Works*, V, 107. See also Black (1803), *Lectures on Chemistry*, I, 521, and Robison (1822), *Mechanical Philosophy*, I, 269–368.

[43] Thomson (1802), I, 368.

The Problem of the Elements

Investigations of early eighteenth-century teaching at the Jardin du Roi in Paris have shown the importance of Stahlian ideas in the lectures given there by the brothers Rouelle.[44] Significantly, their courses were to be singled out for praise by G. F. Venel in his long, brilliant, and influential 1753 article on "Chymie" in the French *Encyclopédie*. Venel's viewpoint may be judged from his statement that the *Nouveau Cours* "brought us Stahlianism, and effected the same revolution in our chemistry that Maupertuis' 'Reflections on attraction' . . . caused in our physics, making us receive Newtonianism." To receive Newtonianism in physics was quite a different thing from receiving it in chemistry. Happily, "this theory which reigns in England, as appears from the chemical works of Mr Hales, has never been adopted here."[45]

In saying this, Venel spoke as the crusader. A passionate, literate, and widely informed defender of the autonomy of chemistry against the very real and obvious barrenness of physicalism, Venel was far from being the sinister obscurantist that some modern commentators, themselves bewitched by physics, have imagined. Venel was well aware of the physicalists and their influence on chemistry. His whole essay was a careful, deliberate, and shrewd attack on the Newtonian course that chemistry seemed in danger of following, even in France. As he quite reasonably observed, the chemical arguments of the physicists "have the great fault of not having been discussed and verified with details and factual examinations. What Boyle, Newton, Keill, Freind, Boerhaave etc. have written on these matters is undeniably marked by a lack of experience." To see chemistry first Cartesian, then corpuscularian, then Newtonian might please some. To read airy physicalist discourses, such as that by John Bernouilli on fermentation, might satisfy others. But to those with the real good of chemistry at heart, the attention paid to such irrelevant and misleading works was tragic at a

---

[44] See Rappaport (1960 and 1961). For a far more extended analysis of the French situation than is possible here, see the 1969 Harvard Ph.D. by M. Fichman, "French Stahlism and eighteenth-century chemistry."

[45] Venel (1753), 437.

time when the writings of Stahl were so little known among both the learned and the fashionable demimonde.[46]

Venel insisted that chemistry must concern itself with the qualities and true interior nature of bodies. Newtonian physics, however useful in its own realm, had nothing to offer here. Its pretended homogeneity of matter, its porosities, and its internal structures could only confuse the chemist. Instead Venel argued "that there are several elements essentially different from each other . . . that the homogeneity of matter is a chimera, that unchangeable bodies, water for example, are composed directly of elements, and that the image of a small structure, under which corpuscularians and Newtonians wish to make us conceive of a particle of water, rests upon ruinous foundations and vicious logic." [47]

If the language was violent, the case was urgent. The need was for chemistry to stand quite apart from physics, and to pursue its own paths. In doing this, it would pay attention to real chemical experiences, such as acidity, inflammability, and color. That chemistry was concerned with the *qualities* and *uses* of bodies was everywhere overlooked in the theoretical schemes of the convinced reductionists. Physics, when taken to the very limit of which humanity was capable, might produce the mathematical principles of Newton. For Venel "the corresponding extreme of chemical genius [is] Stahl's *Specimen Becherianum*." [48]

This was anti-Newtonianism with a vengeance. In Scotland, Cullen was quite aware of how "Monsieur Venelle, who writes le part Chemi dans le Cyclopedia Francois now publishing at Paris, attempts some new regulations in chemistry." To stress principles to the exclusion of Newtonian explanations was to "give in to the Peripatetic system." Cullen preferred simply to

---

[46] *Ibid.*, 408–409. Cf. the article in which Professor C. C. Gillispie hears in Venel's arguments "the authentic voice of the *sans-culotte*" and sees the issue between Venel and Lavoisier as that "between scientists and opponents of modern science": Gillispie (1959), "The Encyclopédie and Jacobin philosophy," 274–275.
[47] Venel (1753), 411.
[48] *Ibid.*, 414.

note Venel's ideas, while making it plain that "I do not agree with him." [49] Cullen might not agree, but the Stahlian system did possess a relationship to chemical reality that escaped the physical theorists. It is thus not surprising that growing chemical industry in France was to prompt a fresh wave of translations of Stahlian metallurgical works in the 1760's, and a fresh awareness of the virtues of Stahl's ideas.[50]

How well Venel mirrored the mood among French chemists may be seen by turning to the far more cautious, but still Stahlian-sympathizing 1749 chemical textbook of P. J. Macquer. Influenced by Maupertuis and also a friend of Buffon, yet uncomfortably aware how physicalist explanations were lacking in chemical utility, Macquer walked a careful line between Newtonian and Stahlian viewpoints. Unlike Cullen, he was content to arrive at earth, air, fire, and water as the four chemical elements. Aware of Newtonian arguments, he admitted how "there be reason to think that these are not the first component parts, or the most simple elements, of matter." Then with Stahlian confidence he went on to argue that

> as we know by experience that our senses cannot possibly discover the principles of which they [the elements] are themselves composed, it seems more reasonable to fix upon them, and consider them as simple homogeneous bodies, and the principles of the rest, than to tire our minds with vain conjectures about the parts or elements of which they may consist; seeing there is no criterion by which we can know whether we have hit upon the truth, or whether the notions we have formed are mere fancies.[51]

Once more we hear the true Newtonian, Cullen, disapprovingly inform his students that "Mr Macquer makes no distinction between chemical and physical elements." [52]

[49] Cullen (1762), I, 17.
[50] Discussed in Guerlac (1959), "French antecedents of chemical revolution."
[51] Quoted from the English translation in Macquer (1758), I, 2.
[52] Cullen (1762), I, 14.

Returning to Venel, we must also note the belief that "the chemist's water is an element, or a substance simple, indivisible and untransmutable, in spite of the sentiment of Thales, Boyle, Van Helmont and Eller." This was of course an unsupported assertion of the most unphysicalist sort. At least it was until Lavoisier so brilliantly strengthened the chemists' hand by directly refuting the work of Van Helmont and Eller and "proving" that whatever the Newtonians might believe, water and earth were *not* transmutable.[53]

A consideration of Lavoisier's background and scientific education is not possible here. So too, the question of his place in relation to the Newtonian tradition is a complex one impossible to discuss at any length. It is however obvious that Lavoisier's great genius lay partly in the ability to be all things to all natural philosophers. Undeniably a believer in the chemistry of principles, he yet collaborated extensively with such a thoroughgoing physicalist as Laplace. In the same way, he avoided positive identification with either the reductionists like D'Alembert, or those like Diderot who favored an organic approach to Nature. He was thus able—no doubt deliberately—to avoid the internecine warfare that characterized French natural philosophy in this period.[54] His unique contribution to chemistry lay in the capacity to synthesize contemporary French currents of Stahlianism, positivism, and natural history, and to use them in explicating the remarkable and puzzling work of the British pneumatists. At the same time he was careful not to challenge directly that Newtonian view of matter common to many of his colleagues and collaborators.

Thanks to the *capture* and chemical recognition of the gases,

---

[53] Venel (1753), 416. That Venel specifically cites the work of Eller suggests new possibilities in the recent discovery of Lavoisier's early manuscript reflections on Eller's theories: see Gough (1968), "Lavoisier—new evidence."

[54] Kiernan (1968), *Science and Enlightenment*, throws some light on the conflicts among the French *philosophes*. Lavoisier's acceptance of a chemistry of principles is well brought out by his most recent biographer: see Daumas (1955). For his close partnership with Laplace, see Professor Guerlac's paper on "Lavoisier and Laplace," to appear in the *Actes* of the XIIe Congrès International d'Histoire des Sciences (held at Paris in August 1968).

The Problem of the Elements

Lavoisier could make the age-long use of the chemical balance a powerful weapon in the argument for his own particular administrative systematics. And the brilliant stroke of inventing a new and orderly language, forced all opponents to negotiate on his terms. Lavoisier's own ability to enjoy the best of all worlds may be seen in his discussion of the chemical elements. Having disproved transmutation, he could afford to mock the physicalists and say in true Stahlian style "that if, by the term elements, we mean to express those simple and indivisible atoms of which matter is composed, it is extremely probable we know nothing at all about them." Yet for all his use of property-bearing principles, Lavoisier was not immune from the Newtonianism of such close comrades as Guyton and Laplace. He was therefore careful also to point out how, "as chemistry advances towards perfection, by dividing and subdividing, it is impossible to say where it is to end; and these things we at present suppose simple may soon be found quite otherwise. All we dare venture to affirm of any substance is, that is must be considered as simple in the present state of our knowledge." [55] Small wonder that Priestley was at times reduced to hopeless rage at the doctrines of his adversary, or that Lavoisier himself did not win the support of his colleagues without a considerable struggle.

In battling against an entrenched, prestigious, and sterile physicalism, great credit must go to that particular strand of French chemical thought which reaches from Stahl through Rouelle, Venel, and Macquer, to Lavoisier. Stahl may have begun the move back toward explanations with chemical utility, but his chemistry was always embarrassed with arguments over the analytical reproducibility of its elements. Not so Lavoisier's. His elements—and above all his oxygen principle—were undeniably the tangible and durable last products of analysis. Even so they were devoid of any satisfactory ontological basis. The subdivision of oxygen was always possible. To this problem Lavoisier had no answer, simply because he refused to make a frontal assault on those Newtonian categories that un-

---

[55] Lavoisier (1790), *Elements of Chemistry*, xxiv and 177.

derlay the physicalist interpretation of chemistry. His failure to discuss the theory of matter, and his highly diplomatic positivism, left his chemistry devoid of a firm philosophical base. By the same token, it preserved him from argument with Newtonian colleagues at a time when his views on combustion, transmutation, and the nature and number of the chemical principles had battles enough to fight.

To deny transmutation and elevate property-carrying principles to a central place in chemical theory was to challenge the whole Newtonian tradition. The challenge was one Lavoisier was careful to avoid making explicit while more urgent problems occupied him. Had his life not been tragically cut off by the guillotine, Lavoisier might well have gone on to tackle those basic questions in the theory of matter which his reformulations of the chemical superstructure so urgently posed. Instead it was left to the English chemist John Dalton, operating in a far different political, theological, and metaphysical context, to provide the new view of matter that would under-pin the reformulation Lavoisier had so forcefully advanced.

7
Quantified Chemistry: the Newtonian Vision

[If there are affinities] which appear to deviate from the ordinary track, they should be considered as comets, of which the orbits cannot yet be determined, because they have not been sufficiently observed.
Torbern Bergman in 1775

7.1. Introduction

The two major problems confronting any philosophy of chemistry have always been to decide on the units that endure through chemical change, and on the terms in which the causes of that change might be discussed. We have seen how mechanistic natural philosophies discounted the first of these problems. By suggesting that the chemical level of organization was less than fundamental, Newtonianism was partly responsible for the doubt, confusion, and argument over the nature and number of the elements, which continued throughout the eighteenth century. The influence of Newtonian theory on discussion of the causes of chemical reaction was more straightforward, though not thereby more fruitful.

How to explain the cause of chemical change was, by the eighteenth century, a problem with an exceedingly long history. One of the most obvious aspects of chemistry is the *specificity* of its reactions. Any self-respecting chemical philosopher of the seventeenth or eighteenth centuries felt it incumbent upon himself to explain why, say, nitric acid and copper reacted vigorously, while nitric acid and gold were devoid of mutual effect. The most natural explanation was that "like assorts with like." This view may be traced through medieval authors such as

Albertus Magnus, back at least to Hippocrates.[1] Seventeenth-century mechanistic chemists were not content with so "Peripatetick" an idea. Instead they suggested that the important thing was not actual similarity, but the possession of particle shapes such that mutual penetration was possible.

An example of the new approach may be found in Lemery's discussion of the solution and precipitation of gold. This discussion makes much use of the sharp edges of particles of *aqua regia*. These edges account for the division of the gold into "subtle parts," and then "[hold] up the gold as if it were like so many finns." Thus solution. Precipitation caused more trouble. Though this was "one of the most difficult [questions] to resolve well, of any in natural philosophy," Lemery was sure that the answer lay in the precipitating agent being able "by its motion and figure" to "engage the acids enough to break them. The particles of gold being left at liberty will [then] precipitate by their own weight." [2]

Such explanations may make us smile, but they were of the essence of pre-Newtonian reductionist chemistry. Their predictive value was small, and their hypothetical character all too apparent in the eighteenth century. Any historian who wishes to see Boyle as "the father of modern chemistry," must bear in mind not only Venel's remarks, but also Cullen's judgment. Cullen argued that "Mr Boyle, Mr Homberg and Mr Lemery, all corpuscularian philosophers, have raised an imaginary system concerning the size and figure of the elementary parts of bodies"; that "the mechanical philosophy though establish'd beyond dispute has yet been misapplied and abused," and that it was folly to deduce the properties of bodies from the supposed sizes and shapes of particles "beyond the utmost reach of our science." [3]

This sober appraisal came from an avowed supporter of the reductionist approach. Venel was less kind. In his eyes, Boyle

---

[1] See *Partington* (1961–64), IV, 569.

[2] Lemery (1698), *Course of Chemistry*, 67 and 69–70.

[3] Manchester University Library, Special Collections, Ms. no. CH C 121 1, pp. 35–36.

could not have a place among the chemists, but only among the physicists. His experiments were careless, his chemical sense lacking, and Becher had long ago pointed out the insufficiency of such things as spiral-shaped air particles. To Venel the moral was clear. Chemistry should avoid the snares of physics and look for a new Paracelsus who would recast and reform the science with clear mind and chemical skill.[4] (It is interesting to speculate on how far Lavoisier, in filling just this role, was influenced by Venel's arguments.)

The Stahlian-oriented among later inquirers might be resistant to any program of numerical measurements of affinities. To many British natural philosophers, and an important Continental group, the lure of a chemistry reduced to the quantified physics of short-range forces was far stronger than the appeal of Stahl. The idea that the chemical force might be quantified like the gravitational force thus came to condition, indeed almost to haunt, much of the research of the latter part of the century. In this period Newtonian astronomy was itself steadily ascending from triumph to still greater triumph. Spectacular progress was being made by the work of a brilliant succession of French mathematicians, from Maupertuis through Clairaut and D'Alembert, to Lagrange and Laplace. In such a context it is not surprising that many chemists, particularly French chemists, should endeavor to make of their own discipline a fully quantitative and predictive science.

It is difficult today fully to appreciate the power, depth, and generality of this ambition in the eighteenth century. On the one hand Newtonian astronomical mechanics no longer possesses either the freshness or the unchallenged command it then enjoyed. On the other, Dalton's successful quantification of chemistry was to be so different in nature and rationale from the earlier vision of the Newtonians that their work has been neglected by commentators ever since. Their work is nonetheless real, and crucial to an understanding of the period. For the greater part of the century, most theoretical argument and

[4] Venel (1753), "Chimie," 416, 435 ff.

experimental effort was based on the widespread and deep-rooted conviction that if chemistry were to be made a predictive, quantitative science, then it would be through the experimental measurement of short-range attractive forces.

A sufficient indication of the great interest in affinity tables which chemists felt throughout the century, and the changing nature of the presuppositions against which they viewed such tables, may be seen in the attitudes of Fontenelle and Fourcroy. In his commentary on the original 1718 table of Geoffroy, Fontenelle had been content to observe that "a chemical table is by itself a pleasing sight, as would be a numerical tabulation arranged according to certain relationships or certain properties." In contrast, Fourcroy—not Keill or Freind or some other Newtonian visionary, but Fourcroy the cautious and positivistically inclined French chemist—was confidently asserting in 1787 that "all chemistry reduces to an exact knowledge of the elective attractions between natural bodies. When the strength of this force between all natural substances has been determined, chemistry will be as complete as it possibly can be." [5]

7.2. The Battle for Quantification

It was one thing to accept the Newtonian belief that the specificity of chemical reaction was to be explained in terms of the differing attractive forces of different substances. It was quite another to measure or even classify these forces. Geoffroy's table was the first to make a comprehensive listing of the preferred reactive orders of a range of chemicals. Like so many of its successors, the table did not allow for the effect on reactivity of temperature, concentration, solubility, or other similar factors. This lack of allowance was not as short-sighted as might be thought. If it is supposed that the ultimate particles of any given chemical have a force of chemical affinity peculiar to themselves, then it is not clear that a change of temperature will necessarily mean a change in the relationship of these forces for

[5] Fontenelle (1718), 37; Fourcroy (1787), *Principes de Chimie,* I, 27.

two chemicals. After all, the gravitational force did not appear to be temperature dependent! Similarly, though concentration might influence the *rate* of chemical reaction, it should not affect the eventual outcome—or so it was thought. The slowly dawning realization that these assumptions were too simple was to be one of the reasons for the widespread disenchantment with the Newtonian vision apparent toward the close of the century.

An additional factor increasing the earlier enthusiasm of practicing chemists for affinity studies was the ease with which such studies could be reconciled with Stahlian as well as Newtonian beliefs. The contrast and potential conflict between Stahlian and physicalist ideas on the elements were very real. Similar conflict was inherent between chemical and matter-and-motion (or "wedge and spiral") explanations of the cause of reaction. But Newtonian beliefs about reaction were more easily reconciled with Stahl's almost animistic forces between different chemical substances.[6] Conversely, Stahlians were unlikely to have the sort of qualms about "attraction" that so bothered orthodox Cartesians.

Though a purist like Venel might object, many practitioners were happy to blend Newtonian and Stahlian ideas on reactivity. Later on, as the lure of force-quantification increased, a split was to develop. But earlier and more qualitative research on tables of affinity could be pursued happily under the banner of either or both persuasions. Indicative is the way that Newton's own writings ensured a wider audience for work of more directly chemical inspiration. His 1706 ordering of the metals in terms of their relative reactivities could in turn be assimilated by Stahl.[7] Quite what the latter thought of the undeniably physical forces of attraction and repulsion in Newton's explanatory commentary is another matter!

The chemical implications of Newton's ideas on attractive forces were quickly grasped and explored in the early British

---

[6] See Metzger (1930), *Newton, Stahl, Boerhaave*, 101–106, 116 and 147–148.
[7] See *Partington* (1961–64), II, 678, and IV, 569.

circle, and widely diffused by the Newtonian popularizers. Even so it is only in the 1750's and 1760's that we find a real coming together of Newtonianism and practical chemistry. Before then the acceptance of affinity studies by chemists was to be seen in the production of a few isolated tables, rather than in a wholehearted commitment to a Newtonian view of matter and its properties. The slow growth of interest in affinity studies was reflected in the number of tables published. Between Geoffroy's original 1718 table and 1750, there were but two new productions. In the 1750's alone there were three, in the 1760's four, and in the 1770's five.[8] Another indication of swelling interest is the appearance of such cautiously approving textbooks as Macquer's 1749 *Elémens de Chymie Théorique*. The increase in avowedly Newtonian chemical texts and research projects from the mid-century was also a function of the growing number of teaching posts available in chemistry.

We have remarked earlier how eighteenth-century chemical investigators of Newtonian persuasion may be roughly divided into medically orientated speculative inquirers, mathematically trained philosophers, and (at least in Britain) members of the popular lecturing tradition. The most significant external influence on the course of chemical teaching and research in Britain was one that developed in the latter part of the century, in close association with this last tradition. The rapidly escalating industrial revolution had as one of its immediate results—a result so far almost totally neglected by historians of science—the demand for chemical instruction, from the new classes of artisans and manufacturers.[9] By the time of Dalton and Davy it had both become possible to obtain one's livelihood through the provision of such teaching, and meaningful to talk of chemistry as a science in itself, rather than as an adjunct to medicine, pharmacy, small-scale metallurgy, or natural philosophy. To see Dalton's atomic theory as an assertion of the independence of

[8] Figures based on the information in Duncan (1962), "Tables of affinity," 178–179.

[9] But see the remarks in a pioneer study by two economic historians: Musson and Robinson (1960), "Science and industry."

the new (industrially related) chemistry from traditional ties and bonds would be farfetched. To neglect the profound influence of the changing nature of chemistry's applications on the professional organization and theoretical formulations of the science would be equally shortsighted.

If we go back to the earlier period, we may note William Cullen's 1747 appointment to a newly created lectureship at Glasgow University as typical of the growing autonomy of chemistry. Cullen's appointment was to influence profoundly the growth of Scottish medicine and manufactures, as well as Newtonian interpretations of chemistry.[10] Again P. J. Macquer, who was to wield an influence in France comparable to Cullen's in Scotland, owed his livelihood for some considerable period to his chair of chemistry even though he, like Cullen, had trained and qualified in the wider field of medicine.

7.2.1. Buffon and the French School

The steps by which Buffon was led to formulate his vision of the universal rule of the $1/r^2$ attractive law, and to outline its implications for chemical research, have been discussed in section 5.5. Buffon commanded great wealth and wide patronage. In France he was able to exert a marked influence on the appointments that were made to the newly emerging class of professional teaching and research scientists. It is thus no accident that the two French chemists who most fully shared his vision also owed much to his backing.

P. J. Macquer,[11] who became *adjoint* in the chemistry section of the Académie Royale des Sciences in 1745, devoted his considerable energies to chemical teaching and research. With the assistance of Baumé, he gave private lecture courses from 1757. In 1771, thanks to Buffon's patronage, he succeeded to the professorship at the Jardin du Roi. His lecturing, his membership of the Académie, and his other official posts ensured him a powerful influence on French chemistry, while the numerous

---

[10] See the essays in Kent (1960), *An Eighteenth-Century Lectureship*.
[11] (1718–1784). See Coleby (1938), *Macquer*.

translations of his textbook and his dictionary created for him a wider audience and a deeper impact. Cullen, for instance, makes repeated references to Macquer's textbook—by no means all of them complimentary, as we have seen.

Macquer's *Elémens de Chimie* was published in three volumes over the years 1749–1751. Its great interest lies in its awareness of and favorable attitude toward both Newtonian and Stahlian ideas. The impact of these ideas on French chemical studies by the mid-century is thus well-displayed and available for study in his work. The contrast with Lemery's *Cours de Chimie* (to be reissued in a fresh edition by the anti-Newtonian T. Baron in 1756), Boerhaave's *Elementa Chemiae*, or Cullen's Scottish lecture course is immediate and striking. Macquer was concerned to give a comprehensive and systematic account of chemical knowledge, from a viewpoint balancing Newtonian and Stahlian themes with a characteristically French stress on natural history. Thus while the break with Lemery's mechanistic position was complete, the exposition was at once more systematic than that of Boerhaave, and far more sympathetic to Stahlian themes than a comparable English work.

Macquer's first chapter, devoted to a discussion of the principles of bodies, was dealt with in section 6.6. His second chapter, entitled "A general view of the relations or affinities between bodies," is also of some considerable interest. Though Shaw had drawn attention to Geoffroy's table, Macquer's was the first elementary chemical text systematically to expound and develop the doctrine of affinities. On the theoretical level, Macquer adopted a cautious and reserved position as one might expect, for Newtonian "attraction" was still a bone of contention among French chemists. (It was after all as late as 1758 that the Académie at Rouen had proposed as its prize subject "to determine the affinities between the principle mixts, in the way that Geoffroy began, and to find a physico-mechanical explanation of these affinities.") [12]

Macquer's caution is seen in his reference to "a mutual con-

---

[12] See Limbourg (1761), *Affinités Chimiques*.

formity, relation, affinity or attraction, if you will call it so." To him "this effect, whatever be its cause, will enable us to account for, and connect together, all the phenomena that chymistry produces." [13] In sounding this note of confidence about the centrality and importance of affinity studies, Macquer was characteristic of his period. Just as the nineteenth century was to place its confidence in its knowledge of the *units* of chemistry and the *structure* of compounds, so the eighteenth saw the charting and explanation of chemical *mechanisms* as its main task. The replacement of the one by the other as the key to chemical understanding might be paraphrased as the move from Newton to Dalton. Certainly those eighteenth-century commentators who were most convinced of the importance of affinities were most ready to see them as Newtonian short-range forces.

Macquer's 1749 discussion of "the nature of this universal attraction of matter" was suitably agnostic. It was noteworthy rather for its careful listing of the different possible cases of simple and double elective attraction, a listing that was to be echoed by subsequent authors. In passing we might also note how Macquer's Stahlian sympathies appear in the statement that "compounds . . . partake of the properties of those substances which serve as their principles." This was but one of several assumptions in Macquer's work, unacceptable to the more thoroughgoing Newtonian chemists in Britain. Of greater acceptability, and more immediate importance to the present study, was Macquer's insistence that "almost all the phenomena, which chymistry exhibits, are deducible from the mutual affinities of different substances." [14]

While the *Elémens* was restrained and agnostic about the cause of affinities, the 1766 *Dictionnaire de Chymie* was more forthcoming. Quite what caused Macquer's change in attitude remains uncertain. Perhaps it was that the later work was orig-

---

[13] Quoted from Macquer (1758), I, 12.
[14] *Ibid.*, 159. See also p. 168 for how Geoffroy's table of affinities is "of exceeding great service."

inally published anonymously, perhaps that a Newtonian approach was unexceptionable in 1766 in a way that was not yet true in 1749. As suggested earlier, it is also likely that acquaintance with Buffon and acceptance of his patronage were two reasons why Macquer firmly embraced a Newtonian position—Buffon's own Newtonian panegyric appeared in that same year of 1766. Whatever the reasons, the sympathies of the *Dictionnaire* were plain to see. And through a variety of editions and translations it was to influence chemical thought and research in a way not possible for the more simple *Elémens*.

At first sight it might appear too bold to call the *Dictionnaire* a work of Newtonian sympathies. Its article AFFINITY stressed that "We do not here enquire into the cause of this great effect, which may itself be considered as the cause of all combinations." Even so, Macquer could not avoid the typically French position that affinities "may perhaps be a property as essential to matter as its extent and impenetrability, of which we can say no more, than that such properties do exist." If this remark was not enough to arouse suspicions of his sympathy with French Newtonians, the clue was soon given in his favorable reference to the works of Newton, Freind, and Keill.[15] The reference put Macquer firmly in the long line of French commentators, from the *Nouveau Cours* through Maupertuis, Clairaut, and Voltaire to Buffon. If we turn to what Macquer has to say of GRAVITY (PESANTEUR), our suspicions are fully confirmed.

It is not possible to enter into Macquer's fascinating discussion as to why the relative density of different integrant particles cannot be deduced from macro-scale measurements of specific gravity, or to dwell on his insistence that the "quantity of matter of any body, is measured by means of any ordinary ballance" and that since "this is the best and justest method to determine the quantity of matters employed . . . it is also the only one which ought to be employed in all the operations

---

[15] Macquer (1766), I, 48. For Macquer's relations with Buffon, see the letters in Buffon (1860), *Correspondance*, I, 139, 150, and II, 63.

Quantified Chemistry

of chemistry, which require accuracy." [16] Instead we must concentrate on his discussion of the cause of chemical affinity. This is a classic in itself. Indeed it so well voiced the hopes, aspirations, and Newtonian vision of late eighteenth-century chemists that Fourcroy was content to reproduce whole sections of it in a work published over four decades later.[17]

Macquer began his exposition by asserting that "the particular effects of gravity much influence all chemical phenomena, and . . . the particular gravity of the integrant parts of different substances is probably the primary cause of the action of these substances upon each other." The similarities between this position and that which Buffon advocated in his *Seconde Vue de la Nature,* also published in 1766, are obvious. Macquer's development of his theme made them even plainer:

> To . . . examine what effects are produced by the gravity of bodies in their combinations and decompositions . . . is undoubtedly the most important and decisive object for the general theory of chemistry . . . Although we do not precisely know the bulks, the masses, the forms, nor the distances of the integrant and constituent parts of bodies, we see them act upon each other, unite or separate, adhere more or less strongly, or resist all union; and we must believe that these effects are produced by the same power; such, for example, as the reciprocal gravitation of these small particles to each other; which is variously modified, according to their size, their density, their figure, their extent, the intimacy of their contact, or the greater or less distance of their approach.

If we leave aside the "Buffonian" belief that the gravitational force itself is also responsible for chemical affinity, there is little in all of this that Keill or Freind, Voltaire or Maupertuis, 'sGravesande or Musschenbroek, Desaguliers or Benjamin Mar-

---

[16] Macquer (1766), II, 185–186. Translation quoted from Macquer (1771), I, 321. Macquer's discussion of specific gravity and integrant particles was omitted from the English translation; but see Macquer (1766), II, 197n.

[17] See Fourcroy (1808), *Encyclopédie Méthodique,* V, 416–422.

tin would have quarreled with. The great significance of Macquer's exposition—so similar in tone and style to Buffon's—is that for the first time a practicing chemist of the first rank was endorsing the Newtonian vision of a quantified chemistry, and doing so in a major chemical text that would have an enduring impact. Macquer's concluding peroration was to leave its readers in no doubt—if doubt they still possessed—as to the importance of this vision. Admitting the superficiality of his own treatment, and the difficulties that lay ahead, he yet felt that "the zeal of persons skilled in mathematics and chemistry" would allow progress to be made. Progress was vital, for such studies were "the key of the most hidden phenomena of chemistry, and consequently of all natural philosophy." [18]

In Guyton de Morveau,[19] Buffon and Macquer were soon to find the man of zeal, "skilled in mathematics and chemistry," that they desired. Guyton first came into contact with his future patron in 1762. As a twenty-five-year-old lawyer and member of the Burgundy parliament, he was then concerned with some legal business of Buffon's. Their friendship ripened, and Buffon did much to encourage and direct his growing interest in science. For chemical instruction Guyton relied on the texts of Macquer and Baumé. (Significantly the *Introduction* to Buffon's *Histoire Naturelle des Minéraux,* produced with Guyton's assistance,[20] was to refer favorably to only three French chemists—Macquer, Baumé, and Guyton himself.) Elected to the Dijon Academy in 1764, Guyton continued his scientific studies. Thanks no doubt to the influence of Buffon, he became a corresponding member of the Académie Royale des Sciences in 1772, the same year that his *Digressions Académiques, ou Essais sur Quelques Sujets de Physique, de Chimie et d'Histoire Naturelle* appeared.

In 1775 Guyton visited Paris and made the acquaintance of

---

[18] Quotations from Macquer (1771), I, 320–328.

[19] (1737–1816). See Bouchard (1938) and the series of articles by Dr. W. A. Smeaton (1957, 1961, 1963, and 1964), on which the author has drawn freely.

[20] See Buffon (1860), *Correspondance,* I, 144–146. See also p. 316, and Pelseneer (1952), "Une lettre de Buffon à Guyton."

Quantified Chemistry

Lavoisier. Back in Dijon in 1776, he embarked on a course of public chemical lectures that were published the following year as the *Elémens de Chymie, Théorique et Pratique*—a title which calls to mind, no doubt deliberately, Macquer's earlier work. We cannot begin to follow Guyton's later, and increasingly successful, career, but must rather concentrate on the content and style of his chemical work as it affects the continuing history of French Newtonianism.

In his recent studies, Dr. Smeaton has characterized Guyton as a "great writer, teacher and critic" rather than as a renowned theorist or experimentalist. While such a valuation appears to underestimate the very considerable impact of his experiments on affinity forces, it if anything increases the importance of his thoroughgoing Newtonianism. Already in the *Digressions Académiques,* his wide reading and physicalist sympathies were displayed in his long discussions of calcination and affinity. And a fresh twist was introduced into Newtonianism, that most flexible of doctrines, when Guyton denied the reality of repulsive forces, preferring rather to talk only of stronger or weaker attractions.[21] The contrast between Guyton's vigorous but unfruitful physicalism, and Lavoisier's own more ambiguous but successful approach, is worthy of remark. At least some of the difference in style and attack may be explained in terms of Buffon's dominant influence on Guyton, as opposed to Lavoisier's more directly chemical and Stahlian heritage through Rouelle, Venel, and the Jardin du Roi.

In 1773 Guyton was launched on the favorite Newtonian topic of capillary rise. G. F. Cigna, a Turin professor, had produced a paper doubting the validity of the Newtonian explanation. The Abbé Rozier, in whose journal the paper was published in 1772, invited the comment of other philosophers. Guyton was quick to oblige, with a paper entitled "Experiments on the attraction or repulsion of water and oily bodies,

---

[21] Guyton (1772), 168–172, 269–293, 302–303. Guyton's Newtonian sympathies were also made plain in his frequent favorable references to Senac, Desaguliers, Musschenbroek, and Freind.

to verify the reliability of the method by which Doctor Taylor estimated the force of adhesion of surfaces, and to determine the action of glass on the mercury of barometers." In this paper, he produced a spirited defense of Newtonian concepts and Taylor's method. The defense resulted in further controversy with F. C. Achard, a Berlin chemist, over points of experimental technique. Guyton also insisted in this paper that "it would without doubt be extremely important to have an exact table of these adhesions for different surfaces"—a remark that was to signal his sustained attempt to use the method which Brook Taylor had pioneered, in order to produce a quantified chemistry.[22]

Guyton's article on affinity in the 1776 *Supplément à l'Encyclopédie* again demonstrated his adherence to the Newtonian tradition transmitted through Freind and Keill ("their efforts have not been totally unfruitful") to Macquer ("one of those who have greatly advanced . . . our knowledge . . . by reformulating and generalizing their theory"). While quoting from Macquer's *Dictionnaire* article on gravity, Guyton did not hesitate to make it plain that "we owe to Monsieur Buffon . . . the route to follow to calculate affinities in the same way as the paths of the stars." His own delight, indeed almost intoxication with Buffon's $1/r^2$ theory is also apparent in the exultant claim that "this beautiful idea opens an immense vista of new knowledge."

Guyton wasted no time in exploiting this opening and pursuing the route that would allow one to "calculate affinities in the same way as the paths of the stars." His rapid success was announced in the 1777 *Elémens*. This work opened with a review of the history of affinity theory, discussing the work of Newton, Taylor, and Desaguliers, the objections of Lagrange and Cigna, and the controversy of Clairaut and Buffon. Of Buffon's importance Guyton was again in no doubt. It was "the Newton of France, the illustrious Buffon, who perceived the

---

[22] Guyton (1773), 172. See also Cigna (1772), Achard (1776), and, for Taylor, section 3.5. above.

veil still covering the truth suspected by the London *philosophe*"; i.e., that the rule of the $1/r^2$ attractive law was universal and all-embracing.[23]

In his extreme enthusiasm for Buffon, and in the zeal of his laboratory inquires, Guyton was atypical. In his belief that chemistry must follow the path of Newtonian astronomy, he was characteristic of the age. From great theorists like Bergman and Berthollet, to enthusiastic popularizers like Fourcroy and Fordyce, all were agreed that a truly scientific chemistry would only be reached by force-quantification. Guyton's own particular contribution was to devise a practical method of quantifying these short-range attractive forces. This he did—adapting Brook Taylor's fir-board experiment—by floating discs of different metals on a mercury surface and measuring the force necessary to lift them clear.

The table of values he obtained was as follows:

Metallic placques of a fixed diameter adhere to mercury with the following forces:

	[grains]
Gold, with a force of	446
Silver	429
Tin	418
Lead	397
Bismuth	372
Zinc	204
Copper	142
Regulus of antimony	126
Iron	115
Cobalt	8

As he triumphantly observed to the world "What is the order that these adhesions follow? It is precisely the order of the chemical affinities." The cause of adhesion was thus experimentally demonstrated to be the same as that of solution and reaction: in each case but one attractive force was at work.

[23] Guyton (1776), I, 183–184, and (1777–78), I, 51.

Atoms and Powers

Unable to contain himself, Guyton went on to rhapsodize about the "satisfying consequences promised by the application of this hypothesis to new observations." Already he could claim exact numerical knowledge of some affinity relationships. For instance, "we are able to say . . . that the affinity of mercury with gold is to the affinity of mercury with zinc as 446:204." Already "one perceives what exactitude these mathematical expressions will bring into chemistry." Naturally the ultimate goal was the Buffonian one. By observing the apparent forces between bodies, and knowing that there was actually only a single $1/r^2$ force at work, one would at length "attain to the rigorous demonstration of the figures which the elements of bodies must necessarily have." [24] Thus, in the moment of success, Guyton uninhibitedly set forth the whole program of research which was to lead to a chemistry completely predictive "like the paths of the stars."

Not surprisingly Guyton was in close correspondence with the Swedish chemist Torbern Bergman. Bergman shared the same Newtonian vision of a quantified chemistry. It was therefore with pleasure that Guyton could announce in the "avertissement" to the second volume of the *Elémens* how "that great law of physics, which makes the stars move, has become in the Uppsala laboratory, as in ours, the key to all the operations that Art demands of Nature." [25] In passing it is amusing to note that Guyton, unlike Venel, saw this drawing together of physicist and chemist in the common task of quantification as the "happy revolution" that lay ahead of chemistry and in contributing to which his own glory lay.

From what has been said, and from the discussion to follow, it will be apparent that the late 1770's and the 1780's were the heroic days of Newtonian chemistry. The long line of speculation and debate deriving from Newton's own work was crystallized in the rival systems of Boscovich and Buffon. The faltering inquiries of Taylor and Hauksbee, and the empirical

[24] Guyton (1777–78), I, 62–67.
[25] *Ibid.*, II, iv–vi.

affinity tables of a variety of chemists were available to experimenters. Thus there seemed to be both a sound theoretical base, and a smattering of technique to guide the Newtonian investigator. With such men as Guyton, Fourcroy, Kirwan, Bergman, and Wenzel engaged in the struggle, progress seemed assured and the quantification of chemical force-mechanisms imminent.

The aim of a quantified science of affinities was most keenly pursued in France. The reasons relate both to the brilliance of French astronomical mechanics at this period, and to the wide influence of Buffon's writings. By comparison, English chemists of the later eighteenth century made but little effort toward the quantification of affinities, despite their almost universal Newtonianism, their belief in the chemical importance of "attraction," and their heritage in the writings of Keill, Freind, and Taylor. This lack of interest in actual experimental quantification of affinity—noticeable in Cullen, Black, and Priestley—is in part to be explained by the bias toward the speculative and empirical rather than the mathematical and rational, characteristic of so much British natural philosophy in this period. It may also be attributed partly to the influence of Boscovich's theory. Certainly the complex force-curve of the *Theoria* scarcely suggested that chemical attraction could be isolated and measured at all easily. In contrast Buffon did possess a simple, overriding vision of the nature and importance of chemical attraction.

In 1774 Buffon was again on the offensive. That he wrote from a theoretical viewpoint rather than as one intimately acquainted with the practical, laboratory problems of chemistry helps to explain his calm and confident faith. Once more he stressed how "the obscurity of chemistry results in great measure from its principles being insufficiently generalized, and not united to the higher physics." In this belief he and his disciples were of course totally remote from the position of Venel. Their vision was also far from the one motivating Lavoisier. To Buffon, Newtonian physics was undeniably the key to chemical advance. Failure to grasp that affinity was but an effect of universal attraction had meant that too many chemists used affinity tables without understanding them. Buffon naturally excepted Macquer and Guyton from this condemnation. Thanks to their work, and his own theoretical arguments, hope lay in the future. Knowing that "all the little laws of chemical affinity which seem so arbitrary" were but examples of the law of attraction common to all matter, provided a "new key" with which "one will be able to explore the most profound secrets of Nature." [26]

The failure of the experimental program and the associated fading of the vision were to change all this, despite Buffon's faith. However if we recapture the atmosphere of the period, we can understand the industry, enthusiasm, and optimism that pervaded the masterly eighty-nine-page exposition of affinity theory which Guyton wrote for the *Encyclopédie Méthodique*. This exposition was published in 1789, the year of Lavoisier's own *Traité Elémentaire de Chimie*. We can thus appreciate why the careful and diplomatic Lavoisier was to admit to "a sentiment of self-love," making him "decline entering upon a work in which [Guyton] is employed." For all his caution, avowed positivism, and strong chemical sympathies, Lavoisier was willing at least to pay lip service to the Newtonian vision. The way he phrased it was to say that "this science of affinities, or elective attractions, holds the same place with regard to the other branches of chemistry, as the higher or transcendental geometry does with respect to the simpler and elementary part": a definite compliment, if curiously two-edged.[27]

The comparative paucity of Lavoisier's own work on affinities has led many commentators to underestimate the place of such studies in the chemical thought of the period. Yet the preface to the *Traité* itself makes plain Lavoisier's own awareness of the powerful Newtonian currents in French chemistry, while also revealing one reason for his concentration in other areas. In this connection it would be interesting to know more of his attitude toward Buffon. Their relationship seems to have been

[26] Buffon (1775), *Histoire Naturelle-Supplément. Tome Premier*, 75–77.
[27] Quotations from Lavoisier (1790), *Elements of Chemistry*, xxi–xxii.

less than total, for even Lavoisier's earlier paper (1783) on the affinities of his newly discovered oxygen made no reference to "the Newton of France." Instead his praise and hopes were reserved for Laplace.

The 1783 paper also serves to illustrate Lavoisier's own mastery of experimental technique, and his superbly logical mind. Pointing out a variety of conceptual weaknesses in affinity tables as then constructed, he concluded that one day it might be possible for the geometrician to calculate the phenomena of chemical reaction in the way the movements of heavenly bodies were calculated. Indeed "the views of Monsieur de La Place on that subject, and the experiments we have projected to explain the affinity forces of different bodies in numerical terms, following his ideas, already permit us to regard this hope as not an absolute chimera." [28] But Laplace was not a man of chemical patience and insight and, whatever his own momentary hopes in the early 1780's, Lavoisier did not proceed with the quantification of chemistry. The enduring sources of his strength were rather in Stahlian thought, natural history, superb experimental ability, and the administrative, ordering, and clarifying zeal of a logical mind reinforced with Condillacean philosophy. An additional commitment to Newtonian reduction was at best a pious hope, at worst a political and administrative device for avoiding trouble in his own French camp.

In this exposition, the stress so far has been on the relationship between Newtonianism and affinity studies. This stress has been deliberate, as we have wished to draw attention to the strength of the French Newtonian tradition, and the importance of the vision of a chemistry quantified like astronomy. This is not of course to deny that affinity investigations were also undertaken by chemists who were far from committed Newtonians. Among those with a more agnostic approach, Baumé, Bucquet, and Fourcroy come to mind. A third possibility was a more militant positivism, such as Lavoisier usually maintained. An extreme form of this positivism may be seen

[28] Lavoisier (1782), "L'affinité du principe oxygine," 534–535.

in the work of Demachy. Though working on affinity tables, he wished to get rid of the word attraction because of its lack of utility and "its danger in the positive sciences" or "sciences of fact." [29]

In this and other ways not only positivists and agnostics but also Stahlians might join at least some part of the great research effort. The closer they were to the Newtonian position, the less the construction of mere rank-ordered affinity lists appealed, compared with the task of numerical quantification and algebraic expression. The vogue for natural history and the sheer experimental difficulties might favor rank-ordered lists. Yet it is evident, particularly from more popular works, that the research effort was nourished primarily by the vision of a chemistry of attractive forces, quantified and predictive like Newtonian astronomy. Among those outside France who shared this vision, the Swedish chemist Torbern Bergman was among the most important.

7.2.2. Torbern Bergman

Born and educated in Sweden, Bergman [30] studied and taught at the University of Uppsala. During his brief lifetime he achieved an international reputation in chemistry, as his *Foreign Correspondence* testifies. Though primarily, he was by no means exclusively a chemist. Indeed his 1758 thesis for the master's degree was a mathematical work. Significantly, its subject was attractive forces. In this thesis Bergman showed that he was already familiar not only with Newton's ideas but also with the work on attractive forces of such men as Keill and Clairaut.[31] It is thus not surprising that Bergman's major lifework, the *Dissertation on Elective Attractions,* was to unite his early interest in Newtonian ideas with his love of chemistry.

---

[29] See e.g., Bucquet (1771), *Etude des Corps Naturels*, I, 7–19; Smeaton (1962), *Fourcroy*, 6–8; Baumé (1778), *Manual of Chemistry*, 3; Demachy (1774), *Receuil des Dissertations*, 292 and 305.

[30] (1735–1784). See Carlid (1955), *Bergman's Correspondence*.

[31] See Bergman (1788–90), *Opuscula*, VI, 44, 46 ff. The thesis was entitled "De attractione universali." For a further example of interest in attraction at Uppsala, see Hiotzeberg (1772), "La cause de l'attraction."

Quantified Chemistry

As early as 1769 Bergman's health was failing, and he retired from active work altogether in 1780, four years before his early death. His great work on elective attractions appeared first in journal form in 1775, then in a more extended version in the 1783 third volume of his *Opuscula Physica et Chemica*. The immediate interest aroused may be seen in Guyton's delighted reception of the first version in 1777, and in his plan to translate the more extended work into French. This latter was also rapidly available in German, and in English translation by Thomas Beddoes. The *Dissertation,* begun in the 1760's when the interest of practicing chemists was still not fully roused, thus became widely available in the mid-1780's. At that time hope was still running high, though the problems of the Newtonian program were becoming more apparent. Bergman's exposition of progress and problems was a model of lucid thought and accurate writing. His careful analysis of the factors involved in any study of affinity, coupled with his exhaustive lists of tables, could not but inspire the faith of Newtonians and encourage them to further work.

Some glimpses of the small and intimate, though international, world of Newtonian chemical research may be obtained from Bergman's *Foreign Correspondence*. This shows that he was in touch with Macquer, and aware of the latter's *Dictionnaire,* from 1768 at the latest. (The *Dictionnaire* was perhaps to serve as his route to Buffon's chemical ideas.) Subsequent letters show Bergman to have been quickly apprised of Kirwan's interest in affinities, and well aware of Guyton's work. We can only conjecture what previous contacts, exchanges of views, and mutual satisfaction lay beneath the iceberg-tip revealed in the 1781 remark to Bergman by a visiting worthy, that "Count Buffon at whose home I spent a fortnight, with Guyton de Morveau . . . has charged me to assure you of his great esteem." [32]

The opening words of the *Dissertation* made it obvious that Bergman was going to treat affinity from a Newtonian point of view. By the second page it was also apparent that he favored

---
[32] Carlid (1955), *Bergman's Correspondence*, 229, 397ff.

Atoms and Powers

Buffon's idea of the gravitational force itself being the force of chemical affinity. "Contiguous attraction" might seem to be regulated by very different laws from that of gravity, but then "the whole difference may perhaps depend on circumstances." The heavenly bodies might reasonably be considered as gravitating points. Not so the intimately mixed particles of chemistry, "for the figure and situation, not of the whole only, but of the parts, produce a great variation in the effect of attraction." Though in this way he favored Buffon's position, the conclusion Bergman drew was a more modest one. He cautiously and correctly argued that "as we are by no means able to ascertain the figure and position of the particles," the most fruitful approach was to "determine the mutual relations of bodies with respect to attraction in each particular case, by experiments properly conducted, and in sufficient number." [33]

Despite the limited nature of his own objectives, Bergman accepted the ruling vision of a quantified science and was quick to admit that "a more accurate measure of each [force], which might be expressed in numbers," was a great desideratum. Footnotes show his awareness that studies to just that end had already begun under Guyton, Achard, and Kirwan. An even more interesting note from the translator of the English version reveals the problems apparent by 1785. This note concerns Guyton's "admirable experiments" and the "very acute and pertinent observation" made by that "frequently superficial" author, Fourcroy. Fourcroy had rightly drawn attention to the danger of amalgamation taking place in experiments with metal discs floating on mercury. In such a case "the difference in the weights, necessary to separate the laminae from the surface of the mercury, may arise, not from any difference of attractive power, but from inequality of mass." [34] So much for Guyton's vaunted numerical results!

Bergman's own work was to avoid such grievous errors, and rightly holds a high place in the history of analytical chemistry.

[33] Bergman (1785), 1–3.
[34] *Ibid.*, 3–4 and 321.

Quantified Chemistry

Indeed the *Dissertation* showed great awareness of, and a lively sympathy for, the problems confronting any investigator of chemical affinities. The discussion and resolution, rather than avoidance, of these problems contributed much to the popularity and reputation of Bergman's work. For instance, he insisted on the need to distinguish between wet and dry reactions. He stressed the need to take account of the effect of heat. He pointed out, systematically and at length, the difference between single elective attraction and double elective attraction (i.e., between $AB + C \to AC + B$ and $AB + CD \to AC + BD$). He was also not afraid to discuss "whether the order of attractions be constant." Needless to say Bergman argued that it was, and that "when once ascertained by experience [it will] serve as a key to unlock the innermost sanctuaries of Nature, and to solve the most difficult problems, whether analytical or synthetical." More emphatically still, "the whole of chemistry" rests upon the doctrine of relative attractions, "at least if we wish to have the science in a rational form." [35]

As the quotations make plain, Bergman fully shared the belief that it was through a study and quantification of force-mechanisms that chemical science was to advance. Particularly revealing of the conditioning power of Newtonian astronomical successes is his remark that should there be a few affinities "which appear to deviate from the ordinary track, they should be considered as comets, of which the orbits cannot yet be determined, because they have not been sufficiently observed." [36]

7.2.3. Wenzel and Richter

Along with Bryan and William Higgins, the Germans Carl Wenzel and Jeremias Richter are most often quoted as "precursors" of Dalton.[37] However, like the Higginses, they are of greater interest for their work on affinities to which, unlike their co-precursors, they made respectable experimental contribu-

---

[35] *Ibid.*, 9.
[36] *Ibid.*, 12.
[37] See e.g., Smith (1856), *Memoir of Dalton*, chaps. 8 and 9; Partington (1937), *Short History*, chap. 8.

tions. Wenzel and Richter are also of interest as being among the few German chemists to respond to the mathematical and Newtonian themes of the "Aufklärung."

C. F. Wenzel [38] learned medicine in Amsterdam and, after a checkered career, became chemist to the Meissen porcelain factory in Saxony in 1786. Quite what prompted his interest in chemical affinity studies remains obscure, but his 1777 *Lehre von der Verwandschaft der Körper* was by no means a negligible piece of work. Now rare, the book was well known in its day, going through three editions. Its interest lies partly in showing how widely known were Buffon's ideas (ideas which Wenzel himself endorsed), and partly in the ingenious method that Wenzel adopted to measure interparticle forces. This method reveals both the difficulties of the subject, and the ingenuity common to its investigators. Wenzel argued that the varying rates at which a given acid dissolved cylinders of different metals should reflect their relative attractive forces. Working on this basis, he produced some plausible sets of figures and excited the admiration of Guyton and Kirwan, if not his fellow countrymen.[39]

J. B. Richter [40] presents a sharp contrast to Wenzel in background, if not in eventual occupation (he too was forced to become chemist to a porcelain factory). Richter studied with Kant at Königsberg, and thus could scarcely fail to be aware of Newtonian thought and the role of mathematics in natural philosophy. Revealingly, his 1789 thesis was *De Usu Matheseos in Chymia*. Convinced that chemistry was a branch of applied mathematics, "he busied himself in finding regularities among the combining proportions where Nature has not provided any." [41] His work exhibited a concern with affinities and num-

---

[38] (1740–1793). See *Partington* (1961–64), III, 671–673.

[39] See Guyton (1786), "Affinité," 586–587; Kirwan (1783), "Experiments on attractive powers," 37. See also Scheele (1786), *Chemical Essays*, 254–259. Both Smith (1856), *Memoir of Dalton*, and *Partington* (1961–64), provide substantial extracts from Wenzel's work.

[40] (1762–1807). See *Partington* (1961–64), III, 674–688.

[41] *Ibid.*, 675.

bers typical of the physicalists of the period (though not of German chemists), but both his obsession with arithmetic and geometric progressions, and his experimental zeal, were unusual. Though known to such fellow inquirers as Kirwan, Richter's work was scarcely of the style to attract a wide audience. E. G. Fischer, a Berlin physics professor, justly observed that this neglect was "the author's fault" for mixing his useful experiments and observations with "calculations which render them obscure to the generality of readers." In undertaking to simplify Richter's results and rescue them from this obscurity, Fischer was unwittingly to provide the first table of chemical equivalents. His aim was however far different. As befitted the period, he was more concerned with obtaining a single table of numbers which might serve "as representing the force of affinity." [42]

The numerical affinity inquiries of Wenzel and Richter were in all probability uncharacteristic of the mood among German chemists of the time. (In this respect it is significant that neither was able to command an academic post.) The dominant currents of thought in eighteenth-century Germany remain as one of the major unexplored subjects of chemical history. Despite our ignorance of their immediate context, Wenzel and Richter demanded mention in this study not only because of their obvious relevance but to redress the balance of previous commentary. Nineteenth-century historians, anxious to establish the "discoverers" of constant composition and to examine "precursors" of Dalton, failed to appreciate the intellectual filiations or contemporary significance of either worker.

7.2.4. *The British Effort*

In France and Scandinavia (and, to some extent, Germany) sustained efforts were made to produce a Newtonian chemistry. Oxford and Cambridge present a sharp contrast. By the second

---

[42] See Berthollet (1803), *Statique Chimique*, I, 134–138. Excerpts from Richter's work are available in both Smith (1856), *Memoir of Dalton*, and *Partington* (1961–64).

half of the eighteenth century they were sunk in inactivity. Richard Watson in Cambridge might show a modest ability at his subject, while Thomas Beddoes [43]—Bergman's translator—had a brief and stormy career as Oxford's chemical lecturer, but in spite of their earlier Newtonian associations, the English universities produced no continuing school of matter-theorists or experimental investigators. However Newtonian chemistry did flourish in Scotland under the able guidance of Cullen and Black. The Irishman Richard Kirwan also made independent but important contributions to the study of chemical attraction.

The early establishment in Scotland of a Newtonian approach to chemistry is apparent from the two published papers of Andrew Plummer, pupil of Boerhaave and predecessor of Cullen in the Edinburgh chair of chemistry.[44] Cullen's own Newtonianism has already been discussed in section 6.5. Cullen appears to have been the first to introduce affinity diagrams with algebraic symbols or numbers to represent relative affinity forces. In his lectures he also made frequent reference to Geoffroy's table. But while he believed in the fundamental role of attraction and repulsion, and stressed that "nothing would be of more importance to the art of chemistry than to establish some just theory" to explain elective attraction, he was also aware that "this has never yet been done." [45]

Cullen's failure to produce chemical publications restricted the range of his influence, a circumstance also true of Joseph

---

[43] On the former, see Coleby (1953), on the latter, Gibbs and Smeaton (1961). Both men deserve further study to aid our understanding of the interactions between scientific activity and political belief within the English Establishment of the period.

[44] (? –1756). *Partington* (1961–64), III, 127–128. See Plummer (1754 a and b): the papers, read in 1738 and 1739, have a thoroughly Newtonian style and content, referring frequently to "attractive powers" and drawing freely on the Queries in *Opticks*. One of the most obvious changes in British Newtonianism as the eighteenth century progressed was the shift of its intellectual "center of gravity" away from Oxford and London to Edinburgh and Glasgow. The causes and consequences of this shift invite examination.

[45] Manchester University Library, Special Collections. Ms. no. CH C 121 1, p. 317. See also Wightman (1955–56), "Cullen and chemistry," 200, and Crosland (1959), "Diagrams as chemical equations."

Black. Black's well-known 1756 paper on *Magnesia Alba* displayed his views on chemical affinity, in passing. Something of the ideas he held late in life may also be deduced from the heavily edited posthumous (1803) version of his lectures. In the *Magnesia Alba*, Black was concerned, *inter alia*, to establish the chemical properties of magnesia. As part of the necessary research, he naturally inquired into its "peculiar degree of attraction for acids, or what was the place due to it in Mr. *Geoffroy*'s table of elective attractions." [46] The vocabulary of "attraction" was freely used throughout his argument, which actually concluded with a series of suggested modifications and additions to Geoffroy's table. The paper thus affords a good view of how Newtonian presuppositions were taken for granted in mid-century Scotland.

The *Lectures on the Elements of Chemistry* are of at least as much use for the study of what John Robison (the editor) felt acceptable in 1803, as for what Black actually said in his lectures. However there is no reason to doubt that the Newtonianism of the *Lectures* accurately reflects Black's own lifelong position. His preliminary castigation of Lemery's "crude and unsatisfactory" ideas on solution was typical of the British revolt against earlier mechanistic theories. Indeed, it could well be said that "we had no chemical theory that connected this science properly with other parts of our knowledge of Nature, until Sir Isaac Newton published that edition of his Optics . . . [with] a number of Queries . . . that relate to chemistry." The principal interest of these Queries was of course the doctrine of attraction, though they also pointed out other paths "in which great reputation has [since] been made by some of the most eminent chemists." In sum, it was not too much to claim that "Sir Isaac's theory, therefore, explains all the most difficult parts of the subject." Indeed to Robison in 1803, if not to Black at earlier times, the theory was "now well established wherever the science of chemistry has made any progress." [47]

---

[46] Quotation from the reproduction in Black (1963), 12.
[47] Black (1803), I, 262, 264–265, and 267.

Though so deeply committed to Newtonian explanations, there is no sign that Black (or Cullen) ever attempted an experimental quantification of chemical attractive force. Thus Black—far from being the great exponent of quantification that he is normally represented as being—was, in terms of the preoccupations and goals of his own period, remarkable largely for his avoidance of the endeavor to quantify. His elegant use of macro-scale weight measurements to follow the course of a chemical reaction might excite nineteenth-century "post-Daltonian" inductivist historians, but there is little evidence that this *method* (as opposed to its *results*) stirred his contemporaries. For, as Black himself evidently agreed, it was in micro-scale attractive forces not macro-scale weights that the key to chemical understanding resided, however useful weights might be for identifying the part played in reaction by such a curious and interesting substance as "fixed air."

Black did little to aid the experimental quantification of chemical forces. He did go in for the weak "number game" form of quantification. This particular approach apparently originated with Cullen in the 1750's. A diagram of Cullen's is shown in Fig. IX and is largely self-explanatory if it is borne in mind that "the numbers placed between the substances express the supposed attractive forces exerted between the substances." [48] The production of many such schemes in the latter part of the century indicates the growing desire for a quantified chemistry. Their use was particularly widespread among British textbook writers, in whom the wish for quantified mechanisms and the obvious didactic utility of such schemes overleaped the need for results deriving from experience.[49]

It seems that the only serious and sustained British attempt at experimental quantification was made by Richard Kirwan.[50] Like Guyton—with whom he became very friendly—Kirwan trained and initially practiced as a lawyer, first in London and

---

[48] Black (1803), *Lectures on Chemistry*, II, 545.
[49] See section 8.2. below.
[50] (1733–1812). See *Partington* (1961–64), III, 660–671.

Quantified Chemistry

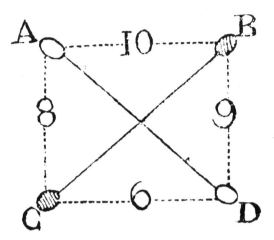

IX. Cullen's "quantification" of chemical mechanism. From Black (1803), I, 544. The reaction considered is AB + CD → AC + BD.

then in Dublin. In 1777 he returned to London. For the following ten years, he devoted himself to chemistry, and it is the research he undertook in this period that is of interest here. During this London interlude, Kirwan was acquainted with the group of gentlemen chemists gathered round Bryan Higgins,[51] a fact which may well relate to that group's faith in quantified affinities.

Kirwan's first major paper on affinity theory was read to the Royal Society in 1780, the year in which he became an F.R.S. Its title was "Experiments and observations on the specific gravities and attractive powers of various saline substances." Its opening paragraph made plain its debts and aspirations:

> The doctrine of chymical affinities hath of late received great improvements from the labours of the very excellent Mr.

[51] (1737–1818). His life and work, and the extent to which he was a "precursor" of Dalton are fully discussed in *Partington* (1961–64), III, 727–736. Gibbs (1965), provides an informative study of the Higgins "circle."

Bergman of Upsal, and the still later researches of Mr. Wentzel; but the order of these attractions has hitherto been the only point attended to by these philosophers, as well as by most preceding chymists: for I know of none, except Mr. Morveau of Dijon, who has thought of ascertaining the various degrees of force of chemical attraction.

Indeed it was Guyton's example that Kirwan sought to emulate, for he had "so ably shown the advantages arising from such an inquiry." [52]

The experimental method on which Kirwan chose to bestow "much pains" again illustrates that, though there was wide agreement about the desirability of the end in view, there was little about the most suitable means of reaching it. Guyton's attempt to obtain a direct measure of adhesive forces was the most straightforward attack on the problem. Wenzel's approach via reaction rates was dependent on the assumption that *rate* of reaction was directly proportional to attractive force. Kirwan's approach was more subtle still. He argued that it should be possible to calculate the specific gravity of a compound from those of its ingredients: "But, in fact, the specific gravity of compounds, found by actual experiment, seldom agrees with that found by calculation, but is often greater." The reason for this increase in specific gravity was "a closer union of the component parts to each other" than occurred in mere mixture. This more intimate union was the result of chemical attraction. Therefore, he "imagined this attraction might be estimated by the increase of density or specific gravity and was proportionable to it." [53] Just like Richter some few years later, Kirwan began to investigate the composition by weight of neutral salts, *in order to obtain a measure of affinity.*

Kirwan had to admit that he "was soon undeceived" as to the validity of his method. Most of his first paper was taken up with discussion of practical problems related to the obtaining of pure

[52] Kirwan (1781), 1.
[53] *Ibid.*, 8–9.

acids and neutral salts, etc., though an attempt was also made to modify the theory to overcome objections. In April 1782 he was back with a long (57 pages) second paper, which was mainly taken up with the correction of practical errors in the first paper. The sheer magnitude of the problems Kirwan faced in obtaining pure materials and consistent results well illustrates the difficulties in the way of chemical quantification at this period—difficulties not mitigated by the long, phlogistic explanations in which Kirwan indulged.

The international nature of the drive to quantify affinities, and the close contacts among the major workers, may again be glimpsed in a letter from the intelligencer J. J. Magellan to Bergman on 26 November 1782. The letter was written two weeks before Kirwan's final paper to the Royal Society. Among other things, it asked for the third volume of Bergman's *Opuscules* (the one that contained the *Dissertation*), if printed, for Kirwan. It also reported that Kirwan was about to give his "third and last" memoir on attractions ("about which he has doubtless already told you in his letters") to the Royal Society.[54] In passing, we may note how the great interest in affinity studies and the close international contacts are also revealed by the appearance of German and French translations of Kirwan's papers—the French translation, significantly, being undertaken in Guyton's home town of Dijon.

In his final paper Kirwan was able to report—as he thought—success. The tables he constructed of acid-base neutralizations have usually been held important by later historians as early examples of the calculation of equivalents. It was not in this way that Kirwan himself viewed them. As he insisted, "the end which of late I had principally in view, was to ascertain and measure the degrees of affinity or attraction that subsist betwixt the mineral acids, and the various bases with which they may be combined." Such an endeavor to quantify affinities was of the greatest importance simply because "it is upon this founda-

[54] Carlid (1965), *Bergman's Correspondence*, 257–258. On Magellan, see Guerlac (1961a), *The Crucial Year*, 36–40.

tion that chymistry, considered as a science, must finally rest." Kirwan went on to discuss the work of Guyton ("his method is incapable of being generalized") and Wenzel ("his method is much more defective") before presenting his own results. These consisted of general rules, tables, and the bold statement that—shades of Guyton—"as these numbers agree with what common experience teaches us concerning the affinity of these acids with their respective bases, they may be considered as adequate expressions of the quantity of that affinity, and I shall in future use them as such." [55]

The criticism of these results by Guyton and Berthollet, Kirwan's own later researches, and his commentary on Richter's work can all only be mentioned in passing.[56] But Kirwan's earlier work, which we have discussed, is of great interest for the way it confirms the widespread desire of chemists in the 1770's and 1780's to make chemistry a truly predictive science through the quantification of force mechanisms. This vision was to be largely obscured in the later 1780's and 1790's through controversy over the phlogiston theory, the new nomenclature, and Lavoisier's insistence on a "chemical revolution." When comparative calm returned to the chemical scene toward the turn of the century, Berthollet was again to take up and pursue the Newtonian vision. His searching critique of the assumptions and methods of previous work, and the vast new program he envisaged, served not to inspire but to discourage fainter hearts.

7.3. The Fading Vision

C. L. Berthollet's [57] clear, logical approach was to reorientate affinity theory, while rendering quite impossible the necessary

[55] Kirwan (1783), 34 and 39.

[56] See Guyton (1786), "Affinité," 592–597; Berthollet (1785), "Observations sur quelques affinités," 305; Kirwan (1791 and 1800), "Neutral salts"—see (1800), 286, for criticism of Richter.

[57] (1748–1822). See Holmes (1962), "Affinities to equilibria," for a study of the post-1803 career of Berthollet's ideas, which complements the information in this present section.

Quantified Chemistry

experimental program. As befitted a pupil of Macquer, he quickly showed an active interest in the study of affinity. In 1785 he was already presenting a paper to the Académie Royale des Sciences, critically discussing Kirwan's work. The assimilation of the new chemistry of gases, together with Lavoisier's reforms and the French Revolution, were to take most of his energy for the next decade. His sustained and radical review of the whole basis of affinity theory therefore came only at the close of the century, partly as a result of his experiences during Napoleon's Egyptian campaign. The 1801 *Recherches sur les Lois de l'Affinité* contained the first statement of his views. They immediately commanded a wide audience. As the translator and epitomizer for the *Philosophical Magazine* was to put it: "Chemists have long been looking for Berthollet's work on this subject, which has at last made its appearance, and contains much new and useful matter." [58]

Berthollet's original treatment was extended, systematized, and made more widely available in his two-volume 1803 *Essai de Statique Chimique* (i.e., essay on chemical forces *in equilibrium*). The opening words of this latter work made clear his loyalty to the research tradition springing from Buffon and Macquer. While "the powers which produce chemical phenomena" are referred to under the name affinity "to distinguish it from astronomical attraction," yet "it is probable that they are the same property." In any case, "it is only since the period that affinity has been recognized as the cause of all combinations, that chemistry could be regarded as a science which began to have general principles." [59]

Berthollet's aim was to clarify these general principles. In particular he argued that "elective affinities do not act like absolute forces," simply because "every substance which has a tendency to enter into combination, acts in the ratio of its affinity *and of its quantity*." The previously ruling suppositions of affinity theory were thus useless. In particular, there was no

[58] *Philosophical Magazine* (1801) *9*, 146–153.
[59] Quoted from Berthollet (1804), *Chemical Statics*, I, vii.

place for "the supposition that elective affinity is an invariable force, and of such a nature, that a body which expels another from its combination, cannot possibly be separated from the same by the body which it eliminated." Instead one had at all times to remember "that the chemical action of bodies . . . does not depend upon their affinity exclusively, but also on their quantity." [60]

The implications of these statements for all previous quantification efforts were immediate and devastating. Berthollet did not hesitate to spell them out. For instance, in order truly to measure the relative affinities of two substances for a third, "it would be necessary to discover in what proportion this third . . . would divide its action." And as he had already pointed out, the snag was "the insurmountable obstacles that would be met with in the means that must necessarily be employed to prove this division of action." [61]

Berthollet's rigorous analysis took in not only the effect of varying reacting masses on the course of the reaction, but also that of heat, solution, cohesive forces, etc. In doing all this he was striving to create the true Newtonian chemical analogue to planetary astronomy. His wish was "to point out, and fix the basis, on which ought to be established the general and particular theories of chemical phenomena." [62] However his analysis served only to illustrate the profound divide between the comparative simplicity and order displayed by astronomical phenomena and the ceaseless variety and change with which chemistry must deal. Far from heralding a new day for affinity studies, his work finally broke the century-long spell of the Newtonian vision. To talk of "insurmountable obstacles" might delight the theoretical purist. It could not enthuse the laboratory researcher. Even Laplace, a lifelong believer in short-range forces, was finally to conclude that the whole program of ex-

---

[60] Quoted from Berthollet (1809), *Laws of Affinity*, 4, 7 ff.
[61] Quoted from *Philosophical Magazine* (1801) *10*, 139.
[62] Berthollet (1809), *Laws of Affinity*, 7.

Quantified Chemistry

plaining chemical affinities by their aid was "useless for the advancement of the sciences." [63]

Perhaps even Berthollet himself experienced moments of nostalgia as he reminded his readers of the simpler world they had lost: "Such was the certainty with which elective affinity has been considered as an uniform force, that celebrated chemists have endeavoured to represent by numbers, the comparative elective affinities of different substances"—but, alas, unhappily without regard to "any difference in the proportion of their quantities." [64] Such innocence would no longer suffice. And without it, tables of numbers and the Newtonian quantification of chemistry (already difficult enough) would prove impossible to achieve.

By curious coincidence 1803 was not only the year that saw Berthollet's *Essai,* and with it the moment of truth for all Newtonian dreams and speculations about quantifying affinity. It was also the year that introduced a far different way of looking at chemistry, atoms, and the road to quantification. That October in England, John Dalton, with considerable surprise, was explaining how "an enquiry into the relative weights of the ultimate particles of bodies is a subject, as far as I know, entirely new." Dalton was able to add that he had "lately been prosecuting this enquiry with remarkable success." The very different traditions and circumstances that led to his work now invite our attention.

[63] Laplace (1878–1912), *Oeuvres,* VI, 392. In characteristic French fashion, Laplace was thinking specifically of Buffon's alluring program for "bringing together under one general law all the phenomena of natural philosophy and astronomy."

[64] Berthollet (1809), *Laws of Affinity,* 4.

8
British Popular Newtonianism and the Birth of the *New System*

> We deal not in those subtilties, where the whole matter of the universe is supposed capable of being compressed into the size of a walnut; or where an inch of common matter may be extended to the size of a world; they serve only to perplex enquiry, and by no means to promote the progress of truth.
> *Adam Walker in 1802*

8.1. *The Popular Lecturing Tradition*

The beginnings of the British tradition of public lectures on natural philosophy may, for our purposes, be located in the coffeehouses of London in the late seventeenth century. The activities of John Harris provide a case in point. Harris was not only a lexicographer, successful author, and Boyle lecturer, but also an intimate of Newton's and sometime secretary of the Royal Society. It was "about 1698, or soon afterwards, that he began to read free public lectures on mathematics at the Marine Coffee House." In keeping with the interests he revealed in the *Lexicon Technicum*, his lectures seem to have been heavily weighted in favor of applied mathematics and natural philosophy. And such was the public interest that the courses, for which a charge was soon instituted, continued until 1707.[1]

The British deliverers of public and vulgarized, or "popular," scientific lectures in the eighteenth century were both inheritors and transformers of the traditions of those theological lecturers who had enjoyed such a vogue in a previous age.[2] Natural philosophy obviously offered a new and secular alternative to religious exhortation. Even so, the links between this philosophy and Christian theology were of continuing impor-

---

[1] See *DNB*, and Taylor (1954), *Mathematical Practitioners*.
[2] On whom, see for instance King (1968), "The Norwich lecturers."

tance within the lecturing tradition (Newtonianism was after all peculiarly susceptible to religious interpretation). The resulting contacts between itinerant lecturers and dissenting academies were of particular value early in the eighteenth century, when industrial progress had not yet dramatically increased the demand for scientific enlightenment.

Restricting ourselves to London for the moment, we may note the public lecture series of such men as the two Francis Hauksbees, James Hodgson, William Whiston, Benjamin Worster, and above all J. T. Desaguliers.[3] We have already mentioned Desaguliers's background as a member of the Christ Church Newtonian group, and the important role he later played as curator of experiments to the Royal Society. This latter appointment, together with his post as Chaplain to the Duke of Chandos, no doubt helped ensure the acceptability in polite society of the technologically oriented lecture courses he delivered. So great was his continuing success that after two decades he could report his "great pleasure" in having "seen the Newtonian philosophy . . . generally received among persons of all ranks and professions, and even the ladies, by the help of experiments." More significantly, he "[couldn't] help boasting—that of the eleven or twelve persons who perform experimental courses at this time in England, and other parts of the world, I have had the honour of having had eight of them for my scholars."[4] Desaguliers's role as trainer of the subsequent generation of popular lecturers was further reinforced by his writings, especially the masterly *Course of Experimental Philosophy* issued in two volumes in 1734 and 1744.

That demand for lecture courses in Newtonianism and its technological implications continued to grow not only in London, but especially in the provinces, is evident from the activi-

---

[3] For details on these men, and a host of other lecturers, see the rich and so-far unexploited repository of information accumulated by Professor E. G. R. Taylor (1954 and 1966), *Mathematical Practitioners*.

[4] Desaguliers (1734–44), *Experimental Philosophy*, I, sig. cl^r. It would be rewarding to identify, and explore the activities of, the "eleven or twelve persons" Desaguliers had in mind.

ties of such diverse figures as Benjamin Martin (initially in Reading), Peter Shaw (in Scarborough), John Horsley (in Newcastle), Caleb Rotheram (in Kendal and Manchester), Matthew Turner (in Warrington), and James Ardern (in Manchester). They in turn transmitted the tradition of lecture courses illustrated by experiments to a later generation of lecturers, including George Adams, James Warltire, "blind Henry Moyes," Adam Walker, Thomas Garnett, and John Banks.[5] Both because of their undoubted importance in aiding the diffusion and transmission of Newtonian ideas, and because John Dalton's earliest scientific ventures were deeply conditioned by the assumptions of later members of this class, their teachings command attention.

Desaguliers himself set the tone, by combining a thoroughgoing Newtonianism with a lively interest in practical mechanics. The possibilities inherent in weights, springs, pulleys, levers, and the laws of hydrostatics were his particular concern, though "Mr. Newcomen's fire engine" also received due place.[6] This whole side of the lecturing repertoire was of course crucial to the developing British industrial revolution, and largely explains the growing eighteenth-century demand for lecture courses. Of more immediate concern here is the view of matter and its properties that the lecturers propagated so widely and successfully. This was at once both definite and orthodoxly Newtonian in the British sense. For instance, we find Desaguliers boldly asserting that "all matter is homogeneous, or of

---

[5] For identification of, and information on, these figures, see Taylor (1954 and 1966), *Mathematical Practitioners;* Gibbs (1961), "Itinerant lecturers"; McKie (1951), "Warltire"; Harrison (1957), "Moyes"; Schofield (1963), *Lunar Society, passim.* The analytic discussion of the number, distribution, training, and importance of the popular lecturers of eighteenth-century Britain has scarcely begun: but see Musson and Robinson (1960), "Science and industry," and Rowbottom (1968), "Teaching experimental philosophy."

[6] That Desaguliers's own interests were primarily technological is apparent both from his textbook and from the startling differences in the experiments he "curated" for the Royal Society when directed by Newton, and when left to his own devices: for the latter, see e.g., *Journal Book of the Royal Society (Copy),* XI, 273; XIII, 92 (on drawing "damps" from coal pits), etc.; for the former, section 3.5. above.

the same nature in all bodies . . . the whole variety of bodies, and the different changes that happen in them, entirely depend upon the situation, distance, magnitude, figure, structure, and cohesion of the parts that compound them." The minute size of the primogenial particles was duly stressed (and accompanied with a veritable battery of supporting illustrations), and the "nut-shell" idea given appropriate mention. Desaguliers constantly assumed that the whole of chemical experience could be reduced to the physics of short-range forces. The limited amount of chemistry actually included leaned heavily on Query 31 in *Opticks*. Characteristic were such statements as "the same bodies that attract one another at certain distances . . . do repel one another at different distances . . . This may be seen upon the dissolution of salts in water. That the parts of salts . . . repel one another at farther distances, appears from the regular figures into which they coalesce, when . . . they are brought within each others sphere of attraction." Or again, "a repelling force is also prov'd by the production of air and vapours." [7]

The doctrines purveyed by the popular lecturers were usually derivative and at some distance from the "research front." Yet too much should not be made of this for, in terms of their period, these men were often among the best-equipped natural philosophers, by virtue both of their knowledge and their extensive experimental apparatus. Even such a vulgar and vastly popular writer and lecturer as Benjamin Martin was quick to learn of and incorporate the new electrical experiments and ether theories of the 1740's.[8] And Desaguliers himself was an important early researcher in this field, as Francis Hauksbee had been before him.

Chemistry attracted less attention from the popular lecturers until the latter part of the century. Even then such chemically oriented men as Warltire, Warwick, and Moyes tended to dismiss the theory of matter with a few brief Newtonian pieties, before turning to the more absorbing problems of practical

---

[7] Desaguliers (1734–44), *Experimental Philosophy*, I, 3 and 17.
[8] See Martin (1747), *Philosophia Britannica*, I, sigs. a3–a4 and p. 11.

chemistry posed by the era's technological progress. Conversely, the more traditional lecturers might dwell on matter and its properties, but without paying too much attention to chemistry. Typical of their style and assumptions was Martin's casual 1751 remark how "by many experiments it appears there is between different kinds of bodies, a different power of attraction . . . indeed all the more considerable operations of chemistry are explicable upon this most simple principle." [9] It was just this sort of remark that Dalton was most likely to encounter when, in the 1780's and 1790's he was struggling to master "mathematics and natural philosophy" for himself.

Though the traditional Newtonian lecturers were as convinced as anyone that chemistry was "physics writ small," their belief lacked compulsion. They had neither the deep concern with forces displayed by those who, in section 4.5., we characterized as the mathematical school, nor the intimate knowledge of chemical phenomena that belonged both to the medical approach and to the newer chemically oriented lecturers. Significantly, it was as a Newtonian lecturer of the traditional kind that Dalton first approached chemistry. Hence, perhaps, his failure to draw on the sophisticated approaches that by 1800 characterized serious Newtonian chemists of both the mathematical and medical schools. Indeed there is little evidence to suggest that Dalton ever appreciated the subtlety and power of Newtonian chemistry, as displayed in such texts as Thomson's 1802 *System*, or Berthollet's 1803 *Essai*.

8.2. Number Games and the Conceptualization of Chemical Composition

While the itinerant lecturers naturally favored a brief and simple discussion of chemistry in its relation to Newtonian matter-theory, more expert chemists were also making their own conceptual simplifications. The known chemical elements might be of no secure ontological status. They might be super-

---

[9] Martin (1751), *Newtonian Philosophy*, 7–10.

seded as knowledge advanced. Nonetheless when Newtonian inspired chemists of the later eighteenth century wished to visualize the process of combination, they did so (often unconsciously) in terms of a one-to-one union of particles of the unknown elements. Passages demonstrating this are not easy to find. The subject of combining proportions was but rarely discussed, being one of little interest. Instead empirically inclined chemists were fully occupied in the vast task of preparing and classifying new substances, while their more theoretic contemporaries were busy with the important endeavor to quantify the forces of chemical affinity. To this end, they might consider the shapes and sizes of chemical particles, and the nature of the chemical power-law, but that elements actually combined one particle to one particle was at once both obvious and unimportant.

It is for instance during a discussion of affinities that Macquer in his *Elémens* casually mentions how, "if a particle of water be intimately united with a particle of earth, the result will be a new compound . . . called a saline substance." More interesting is a passage in the later *Dictionnaire,* where his almost unthinking assumption of one-to-one combination stands fully revealed:

> We may conceive that a neutral salt, for instance, common salt, may be divided into molecules still smaller and smaller, without any separation of the acid and alkali which constitute the salt; so that these molecules, however small, shall always be common salt, and possessed of all its essential properties. If we should now suppose that these molecules are arrived at their utmost degree of smallness, so that each of them shall be composed of one atom of acid and of another atom of alkali, and that they cannot be further divided without a separation of the acid and alkali, then these last molecules are those which Mr. Macquer in his chemical lectures calls *primitive integrant molecules.*[10]

[10] Quotations from Macquer (1758), I, 14, and (1771), I, 28.

Two further British examples may be quickly mentioned. John Elliott's simple and didactic 1782 *Elements of the Branches of Natural Philosophy connected with Medicine* displays (as elementary works so often do) the presuppositions behind more serious inquiries. Elliott was quite certain that "in chemical combination . . . a particle of each element unite together so as to form one particle, considered mechanically." In saying this, he was only expressing more tersely the sentiments of his teacher, Joseph Black. Black said:

> Cinnabar is a compound substance of quick silver & sulphur, and it is an aggregate, as being compos'd of numerous little articles all like one another; & the most minute particles without separating the mercury & sulphur of which it is composed, will be the particles of the aggregate, but each of those is composed of a little atom of mercury & another of sulphur, & these two different substances are the principles of the mixt or compound.

Similar thinking is displayed in William Nicholson's 1795 *Chemical Dictionary*.[11]

Related to the idea of one-to-one combination was the acceptance of constant composition as a characteristic of definite chemical compounds. As far back as 1699, for instance, Homberg's inquiry into the neutralization of acids by bases proceeded on this assumption. So did Henry Cavendish's much later work on the subject. The same belief was also present in Cullen's 1753 statement that "as there are only four species of acids and three of alkali and . . . each acid can be joined with each alkali only in one given proportion, we must imme-

---

[11] Elliott (1782), 104. The Black quotation comes from some 1785 lecture notes cited in McKie (1962b), "Black's lectures—IV," 90. Further examples of the same assumptions are to be found in, for example, Nicholson (1795), *Chemical Dictionary*, 155: "If sulphur and an alkali be combined together, and form liver of sulphur, we may conceive the mass to be divided and subdivided to an extreme degree, until at length the mass consists of merely a particle of brimstone and a particle of alkali. This then is an integrant part."

diately perceive that there can be but twelve species of neutral salts." [12]

Part of Dalton's great insight was to take over these prevailing assumptions, state them explicitly, and then exploit them mercilessly in his pursuit of relative weights. That he ran into charges of plagiarism from one of the few other men to take advantage of these common but unutilized ideas should not surprise us. Dalton made one-to-one combination the starting point of his inquiry into relative particle weights. He thereby inaugurated a new era in chemistry. In contrast William Higgins—the man who later claimed to have "anticipated" Dalton—used one-to-one combination as a basis for a typically Newtonian "number game" with relative affinity forces.

The whole subject of numerical arguments about reaction mechanisms deserves further consideration. Such arguments enjoyed a considerable vogue in Britain toward the turn of the century. Though it was in Germany, Scandinavia, England, and especially France that the quantification of chemical affinities was most actively pursued, it was in Scotland that the explanatory advantages of affinity numbers were first grasped (as was mentioned in section 7.3.4). Already in October 1759 Cullen was writing to George Fordyce, a former pupil, to explain his use of diagrams and algebraic *symbols* to represent double elective attractions. And in 1762 Fordyce was writing back: "I have enclosed two tables [i.e., diagrams] which I propose, provided you give me leave, to prefix to my tables of chemical attraction." There is no firm evidence that Fordyce's intended work ever appeared.[13] However like Bryan Higgins, Fordyce was an influential figure among London's "gentlemen chemists" of the later eighteenth century. The annual course of lectures he gave

---

[12] Quotation from Dobbin (1936), "A Cullen manuscript," 145. And see *Partington* (1961–64), II, 44–46, and Cavendish (1921), *Papers,* II, 66–67.

[13] Thomson (1832), *Life of Black,* 570–571. For a reference to "similar diagrams, published in London, by Fordyce, &c," see p. 146 of Smith (1856), *Memoir of Dalton,* where the ideas on affinity of Cullen, Black, Bergman, Guyton, etc. are discussed sympathetically and at length. See also Crosland (1959), "Diagrams as chemical equations," for a full discussion of Cullen's ideas.

in the capital for thirty years from 1759 made his views widely known, and may well account for the later casual use of affinity force numbers by such chemists as William Higgins.

It seems that Joseph Black was also very quick to use arbitrary numbers to represent affinity forces. Such numbers are found in his lectures as early as 1770.[14] The first *published* use of arbitrary numbers occurred in a work already mentioned, Elliott's 1782 *Branches of Natural Philosophy connected with Medicine*. In this the author acknowledged his debt to Fordyce, and then went on to say, without any preliminary discussion or explanation, that "if vegetable alcali and vitriolic acid attract each other with the force 9, and the nitrous acid, and calx of silver only with the force 2, then if the nitrous acid attracts the vegetable alcali with the force 8, and the calx of silver the vitriolic acid with the force 4; 8 and 4 is greater than 9 and 2: a decomposition therefore will be effected, and two new compounds formed." [15]

Such an unheralded use of affinity numbers suggests that Elliott's readers—Edinburgh students and London lecture audiences—were already familiar with this explanatory device. Just how commonly accepted by the end of the century was such "explanation" of observed reactions by suitably selected forces may be seen in the writings of George Pearson. Reflecting the still small world of the British philosophical chemist, Pearson was both a friend of Fordyce and a pupil of Black. Indeed a lecture syllabus he wrote in 1806 might serve as epitome and epitaph for the whole eighteenth-century Newtonian tradition in Britain: "Chemistry . . . as a department of natural science, according to Newton, Freind, Boerhaave, Lewis, Cullen, Black, Bergman, &c—comprehends the properties of substances which are the effects . . . of the laws of chemical attractions, or affinities." [16]

Another to see the great utility of even arbitrary numbers

[14] Crosland (1959), 82–83.
[15] Elliott (1782), 143.
[16] Pearson (1806), 1–2. See also Pearson (1799).

for explaining the course of chemical reaction was the French chemist, A. F. de Fourcroy. In 1783 he produced an elaborate "account of a new way of explaining the cause of decomposition effected by double affinities through the use of numbers." Though his claim that "no one has perviously adopted the method I propose for explaining double affinities" appears to have been misguided, the belief that "the method will throw much light on a great number of chemical phenomena" was not at all unreasonable. Indeed William Nicholson was to devote many pages of his 1795 *Chemical Dictionary* to diagrams based on Fourcroy's work.[17]

Nicholson was also to serve as translator of Fourcroy's highly successful textbooks. And as he justly observed in his 1804 preface to the eleven-volume translation of the *Systeme des Connaissances Chimiques*: "An entire generation of chemists, who have imbibed their first notions of this science from his works, in every part, and by every language in Europe, can bear testimony to the happy perspicuity, with which he arranges and communicates the important doctrines on which it is founded." When we appreciate that it was this generation of chemists who were to form the audience for Dalton's theory, it will be apparent how important was Fourcroy's acceptance of the fundamental role of affinity theory. Returning to the subject of popularizations, and to Fourcroy's earlier work, we may note how the 1788 translation of his *Leçons Elémentaires d'Histoire Naturelle et de Chimie* also made use of affinity numbers.[18]

It is against such a background of assumed one-to-one combination of particles, and "number-game" arguments about reaction mechanisms, that we must see the work of William Higgins,[19] nephew and pupil to the successful London manufacturing and lecturing chemist Bryan Higgins. In search of

[17] See Fourcroy (1784), *Mémoires et Observations*, 308–323, and Nicholson (1795), I, 173–187.
[18] Fourcroy (1804), I, v: (1788), I, 71 *et seq.*
[19] (1763–1825). See Partington and Wheeler (1960), *Life and Work of Higgins*, for an exhaustive discussion of Higgins' work, and a facsimile reprint of his two books.

weapons to use against the phlogiston theory, William hit on the idea of using completely arbitrary affinity numbers to reinforce his arguments. It is this that makes his 1789 *Comparative View of the Phlogistic and Antiphlogistic Theories* an interesting and amusing, if unconvincing, book. The style and assumptions of its arguments may be seen in the following passage:

> Let S be an ultimate particle of sulphur, recently deprived of its dephlogisticated air, and still possessed of the power of $6\tfrac{7}{8}$ to recover this again; and let $S \xrightarrow{5-1/18\ d}_{5-1/18} D$ be a particle of vitriolic acid in the vicinity of S: will not S take D or d from S [of the vitriolic acid]? and will not the volatile compounds $S \xrightarrow{6-7/8} d$ $S \xrightarrow{6-7/8} D$ be formed? The latter will pass off in an elastic state.[20]

Not surprisingly, the passage does contain among its unstated assumptions ideas on combining proportions that Dalton was to make explicit. Yet the *Comparative View* is important chiefly as a brilliant and highly individualistic exploitation of the dominant Newtonian assumptions about affinity forces, rather than as some pale precursor of Dalton's later work. Indeed Higgins' book only makes more acute the questions as to how and why Dalton was so powerfully to transform and redeploy these assumptions. Before these questions can be broached, one final element in the background of prevailing belief demands investigation.

8.3. Matter-Theory and Theology

That theories of matter, and of the relationship between brute matter and the "rational soul," should be of great theological consequence in the eighteenth century is at once both obvious and unexplored. In earlier sections, we have sought to sketch

---

[20] Higgins (1791), 44.

something of the theological preoccupations that lay behind Newton's thought. We have dwelt on his maturing scheme of gravity as an immaterial power, operating in the almost totally vacuous universe envisioned by the "nut-shell" theory. Such a scheme was of obvious appeal for the way it combatted the materialistic atheism so easily associated with "corpuscular philosophy" by both followers and opponents of Descartes, Hobbes, and Spinoza.

Newton's own particular unease over "brute matter" led to the "nut-shell" theory: a theory powerfully related to the experimental observations of the *Opticks*, if philosophically less satisfactory. It is not surprising that a theologian well-aware of the Newtonian position soon developed a philosophically more coherent, though logically more extreme, attempt to defend Christianity from the perils of materialism. In 1706 George Berkeley was already agreeing that "matter tho' allowed to exist may be no greater than a pin's head." His notebooks show him correspondingly aware of "Keill's filling the world with a mite." Berkeley himself was less than impressed with this idea so popular in the Newtonian group, for he saw "no wit in any of them but Newton. The rest are meer triflers, mere Nihilarians." [21]

Berkeley's own radical idealistic theories cannot concern us here, except to note their Newtonian filiations, their characteristically British concern with the relationship of matter-theory to theology, and their profound influence on the course of future debate. This influence was most plainly apparent in the reaction Berkeley's ideas provoked. The whole of Scottish "common-sense" philosophy may be seen as a robust response to those deepening doubts about the reliability of our knowledge of the external world which, growing through the work of Locke and Berkeley, found such alarming expression in the ideas of David Hume.[22] And because common-sense philosophy was both opti-

[21] Johnston (1930), *Berkeley's Commonplace Book*, 13 and 43.
[22] See Stephen (1876), *Eighteenth-Century English Thought*, especially chap. 1. Among its many virtues, Stephen's study is one of the few works to give any consideration to the relationship of matter-theory and theology in this period.

mistic about the reliability of sensory knowledge and widely influential among the natural philosophers associated with the Scottish universities, it in turn was to have an important and so far unacknowledged influence on the course of science late in the eighteenth century.

Berkeley's idealism represented one path open to devout Anglican thinkers reacting against the possible materialism of even Newton's work. Other roads were also pursued. In an earlier section (5.2), Robert Green's work was discussed. His distaste for Newton, and for those "of a much lower class, Mr. Hobs Mr. Lock and Spinoza," led to a system in which forces were paramount, and "matter" in the usual sense was banished. Equally radical in a different way was the scheme propounded in John Hutchinson's *Moses Principia* of 1724, and a host of related works. Hutchinson's ideas commanded a considerable following among more conservative Anglicans throughout the eighteenth century. They gained their most coherent exposition in the writings of the cleric William Jones, notably in his 1781 *Physiological Disquisitions*. And the growing strength of Evangelicalism, coupled with the great resurgence of conservative thought in all its forms, gave these ideas a fresh strength and topicality in England in the 1790's.[23]

Hutchinson desired above all to maintain the supremacy of the biblical record. To this end he advocated a new system of reading Hebrew "without points." At the same time he was concerned to deny not the mathematical accuracy but the physical validity of a Newtonian system based on immaterial forces. He argued that the first chapter of Genesis clearly showed the existence of an ethereal medium, composed of fire, light, and

---

[23] Thus Hutchinson's major work was published in 1724, saw a new lease of life (especially among Oxford's proto-Methodists) through his *Collected Works* (1748), and Horne (1753), and was later influential via the writings of William Jones (1762 and 1781) and the 1799 *Collected Works* of Bishop Horne. In this study, it will not be possible to do more than refer glancingly to the Hutchinsonians, one of the most intriguing of eighteenth-century groups, who deserve more than Stephen's slighting reference to how "the crotchets of weak minds may indicate the general current of speculation." See also Metzger (1938), *Attraction Universelle*, 197–200, and Kuhn (1961), "Glory or gravity."

air. The value of such a medium to Hutchinson was that it helped avoid the danger of pantheism, for its existence allowed a clear demarcation of the material from the spiritual world. It thus helped defend the faithful from such dangerous delusions as that thinking, like attraction, was simply a property of matter. "Gross matter" once again became solid, passive, and inert. Gravitational attraction and all other forces were not themselves *causes,* but merely impact-results of ether mechanisms. Inactive, solid, and helpless, matter was thus once more sharply distinguished from all things spiritual.

Not surprisingly, Hutchinsonian ideas enjoyed a vogue for rather different reasons towards the mid-century "when the surprising phenomena of electricity" seemed to demand an ethereal explanation. These phenomena, "which seem greatly to have awakened and extended the curiosity and attention of mankind," could naturally be interpreted as powerful support for Hutchinson's system.[24] And later still this same system could be used to counter not the pantheism that had earlier stalked immaterial forces but—paradoxically—the materialistic atheism these same forces were now thought to encourage.

By the 1770's immaterial forces were no longer so obviously a proof of God's power, even in Britain. Consequently "a *vacuum,* that basis of all philosophy and barrier against atheism," was no longer "the sponge of all atheistical systems" [25] that it had been for the zealous Newtonians of the early eighteenth century. Indeed, as the century progressed, the deistic and free-thinking associations of Newtonianism became more obvious to many British commentators than its original orthodoxy. The example set by the French *philosophes* in the use of Newton was too close to ignore. And the writings of Joseph Priestley brought home the threat to any of the devout who were still complacent. Already at an earlier stage, the tenor of thought among the pietistic, conservative, and evangelical may

[24] Horne (1753), *The Case between Newton and Hutchinson*, 3. Awakening interest in electricity was here in happy conjunction with incipient Methodism in promoting Hutchinson's ideas.
[25] *Ibid.,* 47.

be seen in the writings of George Campbell, Isaac Watts, and John Wesley. Typical was Wesley's sympathy for the Hutchinsonians rather than mainstream Newtonians: not Pemberton's *View of Newton's Philosophy* but the *Works* of John Hutchinson was the reading he recommended to students.[26]

While Hutchinsonianism provided an important element in the British pattern of thought on matter-theory, the line of attack pursued by Berkeley led rather to David Hume. Hume's devastating clarity and sceptical intent, if never given the measure of understanding they deserved, were recognized in the later eighteenth century as threats to religion and public morality that must be met. The most influential answer was the one slowly developed by Thomas Reid, James Beattie, Dugald Stewart, and other philosophers of the Scottish "common-sense" school. They argued for both the reliability (if not comprehensibility) of sense experience and the reality of brute matter. To them the sophistry of Berkeley and the more dangerous, though muddle-headed, delusions of Priestley were equally unacceptable. Typical of the earthy realism of the position they developed were Reid's comments on Berkeley: "To me nothing seems more absurd, than that there should be extension without anything extended; or motion without anything moved: yet I cannot give reasons for my opinion, because it seems to me self-evident, and an immediate dictate of my nature."[27] Such arguments as this undoubtedly helped prepare that climate of opinion into which Dalton's highly visual and "realistic" theory of solid and impenetrable chemical atoms could be born. The hostile British reaction to Priestley's work was a further important factor in creating that climate.

In exploring the tangled web of relationships between chemistry, matter-theory, and theology in the late eighteenth century, a special place must be given to Joseph Priestley. This is not

---

[26] See Schofield (1953), "John Wesley and Science," 339.

[27] Reid (1785), *Intellectual Powers*, 258. For Reid's (unpublished) reply to Priestley's *Disquisitions*, see McCosh (1875), *The Scottish Philosophy*, 473. For Priestley on Reid, see Priestley (1774), *Examination of Dr. Reid's Inquiry*.

because his ideas, ingenious though they were, displayed a dazzling originality. It is rather because the polemical writings of this mildest of men stirred hostility, rage, and reaction in a way not true for any other author. We have already noted (in sections 3.4.1 and 6.5) the steady progression in ideas from Newton to Boscovich, and how Priestley took this whole development to its logical conclusion in his resolution of the matter-spirit dichotomy by the calm abandonment of solid matter. In Priestley's work, this abandonment was intimately associated with the denial of immortality and free will, and the adoption of the doctrine of "philosophical necessity." This was strong meat. That it was far too strong for the orthodox is immediately plain from Boscovich's heartfelt cry that Priestley's matterless universe was one of "materialism pure and simple."

Disquisitions Relating to Matter and Spirit was published in 1777. The flurry of pamphlets that followed bears ample witness to the fact that Priestley's ideas were received with alarm.[28] It was not only Continental Jesuits and Scottish philosophers who objected. Even his friend Richard Price was concerned to say that Priestley "ought not to lay so much stress as he does on the doctrine of materialism." Of course, materialism here meant the denial of the matter-spirit dichotomy. Characteristically and paradoxically, the wish to preserve this dichotomy led Price to re-emphasize how matter "must consist of solid particles or atoms occupying a certain portion of space, and therefore extended, but at the same time simple and uncompounded . . . This seems to have been Sir Isaac Newton's idea of matter." John Whitehead, an evangelically inclined Quaker, voiced similar sentiments, maintaining the conservative position that "solidity, impenetrability and vis inertiae . . . are . . . properties essential to [matter's] being or existence." He also gave utterance to the common fear that "materialism"—by which of

---

[28] See Crooke, *Priestley Bibliography*, sec. 4. A glance at the *Bibliography* immediately reveals how peripheral to Priestley's main concerns was natural philosophy. Theology and politics take 102 pages, science, 21.

course he meant the denial of solid atoms—"must terminate in atheism." [29]

If Priestley's ideas were viewed with unease in the 1780's, that unease was as nothing compared with the hostility of the 1790's. The giddy course of the French Revolution seemed to fulfill every British prophecy of the consequences of materialism and atheism. Priestley's personal popularity in France served only to reinforce the xenophobia so apparent in England. That the course of political events should have thus served to influence the reception of Priestley's ideas on the nature of matter is natural enough. That one of the most sustained denigrations of those ideas occurred in the earliest work to praise John Dalton illuminates the complex roots of chemical atomism.

In the turbulence of the 1790's, it was predictable that many British writers on natural philosophy would wish to distinguish the object of their studies from that notorious "philosophy" so associated with French troubles. One of the more systematic attempts to do so was that of George Adams,[30] a popular lecturer, in his 1794 *Elements of Natural Philosophy*.

Adams made no bones about his alarm. In France "principles were . . . propagated under the veil of philosophy, that are subversive of all order and religion." Worse still "it was publicly avowed, that the men who were pursuing here the schemes that have made France a scene of ruin and desolation, were known to be philosophers." [31] With this veiled reference to Priestley among others, Adams set about rescuing science from such company. The method chosen was to expound an extremely conservative variant on Newtonian philosophy, while mounting oblique attacks on Priestley's ideas (Priestley was named only in the index, where the appropriate fourteen-page assault was listed

---

[29] For Price's remarks, see Priestley (1778), *Free Discussion of Materialism*, 339 and 10–11. See also Whitehead (1778), *Materialism Philosophically Examined*. The quoted passages are taken from the review in *Gentleman's Magazine* (1778) *48*, 381, and from the reply in Priestley (1778), *Free Discussion of Materialism*, 198.

[30] (1750–1795). See *DNB*.

[31] See Adams (1794), I, viii–x.

as "Priestley, Dr. . . . His system of materialism fully examined and confuted").[32]

Adams moved in circles closely related to those of Dalton. In his book, we can observe at work many of the forces by which Dalton was also influenced: the appeal to common sense, the conservative theological bias, the stress on the utilitarian, and the reaction against infinite divisibility, the "nut-shell" theory, and anything tainted by association with Priestley. Thus Adams made no bones about his gratitude to such a prominent Hutchinsonian as "the Rev Mr. William Jones [to whom] I am particularly indebted." Like Jones, he explicitly disavowed the infinite divisibility of matter, stressing instead "the necessity of supposing all matter made up of atoms," each of which "occupies its own space" so that "nothing else can be said to exist where it is." As befitted his conservative stance, Adams was naturally convinced that matter of itself was passive and inert. Attraction was not inherent, but to be explained by fluid mechanisms. And, most interesting of all, Adams chose to break with the standard Newtonian belief in the inertial homogeneity of matter. His argument that gold being sixteen times as (inertially) dense as water does not allow us to decide anything about their relative quantities of matter may well reflect his wish to escape the snares of the "nut-shell" theory: what is of particular interest here is how closely his muddled arguments prefigure Dalton's later attempts to defend heterogeneous matter and chemical atoms.[33]

The interrelations of political reaction, theological conserva-

[32] *Ibid.*, V, 33.

[33] *Ibid.*, III, 48–49. A denial of inertial homogeneity seems to lurk in the works of the Hutchinsonians. See e.g., Jones (1761), 29–38. Cf. Jones (1781), 6–26. Consider also how closely Jones's ideas resemble those Dalton was to advocate: "[It is] false, that all the matter in the world is homogeneous: for there is undoubtedly an essential and radical distinction between the elements of which bodies are composed; so that neither artifice nor violence can ever transmute earth into water or lead into gold." Thus, for very different reasons, the Hutchinsonians agreed with the French chemists in advocating immutable elements. Dalton's awareness of the former group may well explain why he—unlike Davy—saw no difficulty in accepting the immutable elements of the latter group.

tism, common-sense philosophy and matter-theory are obviously too subtle and involved for any clear and unambiguous correlations to be made. What we have rather sought to achieve in this and previous sections is to draw attention to some neglected features of John Dalton's background. These features will help us to understand more adequately the form taken by his work. They also illustrate the ways in which the development of British natural philosophy was a function of the changing social as well as intellectual context. As we appreciate more of Dalton's Quaker piety and Scottish contacts, of his lecturing friends and mathematical interests, and of his manifest unease with much Manchester radicalism, so we may the more adequately comprehend that range of social and intellectual factors shaping his *New System of Chemical Philosophy*.

8.4. *John Dalton's Scientific Education*

To discuss John Dalton's [34] scientific education at any length is not possible here. Yet because the subject is one so confused by half-truth and confidently repeated error, a brief clarification is needed.

Dalton was born in the agricultural and mining area of West Cumberland in 1766. He was the younger son of a Quaker smallholder and handloom-weaver. The dominance of Quakerism in the area was crucial to his intellectual development, for the sect's contemporary emphases on pietistic and conservative theology, on education, natural philosophy, and individual enterprise were to leave deep marks on the course of his career.[35] Dalton has often been viewed as naive, self-taught, and virtually unlettered in science before his 1793 removal to Manchester. Such views are mistaken. They spring from the twin errors of taking too literally his own later disclaimers, and un-

---

[34] (1766–1844). See Greenaway (1966), *Dalton,* and Cardwell (1968), *Dalton and Science.*

[35] On Quakers and the widespread pietism of the late eighteenth century, see Jones (1921), *Later Quakerism, passim.* For Dalton's deliberate identification with the traditionalists (by his dress and speech), see e.g., Lonsdale (1874), *Dalton.*

critically accepting the posthumous verdict delivered by his adopted city during its own Victorian heyday. In fact Dalton was quickly enmeshed in that Quaker network of natural philosophers which covered the English-speaking world of the later eighteenth century.[36]

If his father's fecklessness and his own position as younger son militated against professional training, his natural abilities were quickly recognized by friends and relations. "Cousin" Elihu Robinson soon took Dalton in tow. Robinson was himself widely read and traveled, and of considerable standing among Cumberland Quakers. His friends included such nationally eminent men as the physician John Fothergill, while his own amateur scientific pursuits embraced at least meteorology and natural history.[37] It is thus understandable how at the age of thirteen Dalton was already busy copying out verbatim such a popular annual almanac of Newtonian mechanics, astronomy, and meteorology as the *Ladies Diary*.[38]

An obvious mathematical and pedagogical ability soon combined with economic necessity to cause Dalton to move, when barely fifteen, to an assistant's post in the Kendal boarding school run by yet another Quaker cousin, George Bewley. Once again, an apparently unpromising exterior hid a peculiarly favorable situation. The school to which Dalton moved was newly built and equipped by the Quakers. The subscription list was headed by Fothergill and supported by such wealthy Midland entrepreneurs as Abraham Darby and Richard Reynolds. More immediately important than the web of contacts this list reveals is the use that the school's first principal made

---

[36] The present author hopes in due course to publish a more extensive study of Dalton's early development within his Quaker context, based on still-surviving and previously unexplored manuscripts in Cumberland and Westmorland. The Victorian historiography of Dalton's life and work is analyzed in an introduction to a reprint of Roscoe and Harden (1896), *New View of Dalton*, now in press for the Sources of Science Series. The account of Dalton's scientific education that follows should be supplemented by reference to Thackray (1966c), "The emergence of Dalton's theory."

[37] See the thirty-three letters to and from Robinson in the Library of the Society of Friends, London. See also Wigham (1890), "Old letters."

[38] Lonsdale (1874), *Dalton*, 39.

Atoms and Powers

of the £150 available for the library. George Bewley was quick to purchase not only Newton's *Principia,* but also the supporting texts of 'sGravesande, Pemberton, and Rutherforth, to say nothing of such works as "Halley's Astronomical Tables" and "Maupertuis' Figure of the Earth." Later purchases included "Musschenbroek's Natural Philosophy," "Boyle's Works 6 vols," "Buffons's Natural History," and "Flamstead's Historia Coelestis." The collection was rounded off with various items of apparatus, including "a two foot reflecting telescope," "a double microscope," and (for £21-0-0) "a double barrel'd air pump with apparatus." [39]

It is significant that Dalton did not feel such resources as these worth even a mention in the accounts of his early life that he was later to authorize for publication. Nor did he refer to the stimulus available to such a talented and enterprising youth from the lectures of itinerant natural philosophers. These lectures must have served to supplement the Newtonian mathematical and mechanical theory so freely available to Dalton in the texts of the school library. Typical of the courses offered was that of John Banks in 1782. During a stay at Kendal of at least seven weeks, he gave "twelve lectures, which include the most useful, interesting and popular parts of philosophy." The lectures were illustrated by extensive apparatus. How closely Dalton must have attended such courses was apparent when in 1787 he in turn offered the residents of Kendal a series of twelve lectures of his own embracing mechanics, optics, meteorology, and "the use of the globes." [40]

The scientific, mathematical, and Newtonian riches available via George Bewley and the school, and vividly displayed in the courses of visiting lecturers, were all to seem as nothing in retrospect. Not so the presumably still greater resources offered by the library, knowledge, and enthusiasm of another Kendal

---

[39] Friends' Meeting House, Kendal. Mss. Packet no. 98; 4to volume on "Friends School House, Kendal." See pp. 1–12.

[40] See *Cumberland Pacquet,* 5 Nov. 1782, p. 3. Banks described himself as "Lecturer in Philosophy" and spent almost a year in a tour of Cumberland, Westmorland, and the Isle of Man. See also Harden (1915), "John Dalton's lectures."

Quaker—John Gough,⁴¹ the blind natural philosopher of Wordsworth's "Excursion." As early as 1788, Dalton was enthusing how

> John Gough is . . . a perfect master of the Latin, Greek, and French tongues . . . under his tuition, I have since acquired a good knowledge of them. He understands well all the different branches of mathematics . . . He is a good proficient in astronomy, chemistry, medicine, &c. . . . He has the advantage of all the books he has a mind for . . . He and I have been for a long time very intimate; as our pursuits are common—viz. mathematical and philosophical—we find it very agreeable to communicate our sentiments to each other, and to converse on those topics.⁴²

Gough's influence on Dalton may be seen in the latter's development of a love for meteorology that was to prove lifelong, and one for botany (Gough's specialty) which, though ardent, was short-lived.

It is clear that while Dalton enjoyed neither the rigorous mathematical training of a Boscovich or a Laplace, nor yet the Newtonian oriented medical background of a Cullen or a Black, it would be a mistake to suppose him scientifically unaware or untutored in his youth. Indeed he was actively linked to the theological, lecturing, and pedagogic traditions of British popular Newtonianism throughout the crucial years of adolescence and early manhood. Significantly, his taste was first for "things mathematical," then those "philosophical." His exposure to chemistry was slight, and his ideas on the nature and properties of matter were formulated in a theologically conservative context.

From as early as 1783, before his seventeenth birthday, Dalton's name had begun to appear among those of correspondents successfully answering the mathematical puzzles in the *Gentle-*

---

⁴¹ (1757–1825). See the account in *Westmorland Gazette*, 7 June 1884, and Nicholson (1861), *Annals of Kendal*, 355–368.

⁴² Henry (1854), *Life and Researches of Dalton*, 9–10. See also Dalton (1834), *Meteorology*, xvii–xviii, for a public acknowledgment of Gough's influence.

man's Diary, and from 1784 for its sister publication, the *Ladies' Diary.* The earliest actual printing of one of his solutions occured in 1789, in the *Ladies' Diary,* which characterized him—no doubt according to his own description—as "teacher of the mathematics in Kendal." Such was to be the spur of this success and his ambition that by 1790 Dalton had simultaneously in print one answer in the *Ladies' Diary,* two in its *Supplement,* five further mentions for successful solutions, two answers in the *Gentleman's Diary,* and two further questions of his own devising! [43] By this time too his contributions were ranging over questions of morals and meteorology as well as mathematics, and their variety and profusion must have served to make his name known in minor scientific circles, and to aid his 1793 move to Manchester.

8.5. *The Origins of the Chemical Atomic Theory*

When Dalton achieved the promotion he so actively desired, and became Professor of Mathematics and Natural Philosophy in Manchester's recently established dissenting academy, his first scientific book was already with the printer. His *Meteorological Observations and Essays* made little immediate impression on the larger national or international worlds of science. The perceptive reporting and pregnant suggestions it contained (suggestions that formed the essential base for the later chemical atomic theory) were generally ignored. It is therefore the more significant that the *Meteorological Observations* received rapid and enthusiastic endorsement in George Adams' 1794 *Elements of Natural Philosophy.* This reception would seem indicative of Dalton's standing in the worlds of both popular lecturers and conservative matter-theorists.[44]

---

[43] For a bibliography of Dalton's writings (published and unpublished), see Smyth (1966); for an account of these early papers, Wilkinson (1855).

[44] See Adams (1794), *Natural Philosophy,* IV, 540, 560–561 (and see the remarks on Adams in section 8.3 above). It seems reasonable to conjecture that Dalton was also aware of the ideas of Adam Walker (himself a native of the Lake District, sometime Manchester resident, and popular lecturer) as expressed in Walker (1802), *Familiar Philosophy*: see the quotation from this work at the head of the present chapter.

British Popular Newtonianism

Further hints as to Dalton's filiations, and his unease over radical theology and philosophy in its Unitarian form, may be gathered from his relations with his employers. That Dalton should have joined the Manchester Academy at all is surprising when seen apart from his ambition for life in a more scientifically active center than Kendal. That his position within such a hotbed of advanced thought was always peripheral and uncertain is apparent from the Managers' minutes.[45] In fact Dalton left the Academy and became an independent teacher just as soon as he was able.

It was as a professor in the Academy that Dalton first began to pay serious attention to chemistry. He did so as an operational necessity, being instructed to deliver courses in the science.[46] The chemistry he thus taught himself was technologically directed. It was also almost unconsciously grafted on to his earlier knowledge of mechanics, the theory of matter, meteorology, and natural history. From the start his chemical ideas were thus to be under the dominance of that particular view of matter, and that highly mechanical view of chemical processes, already apparent in his first book.

Dalton's earliest meteorological researches had not unnaturally awakened a deep and abiding interest in the state of water vapor in the atmosphere, and the theory of rain. The 1793 *Meteorological Observations* even went so far as to advance "a theory of the state of vapour in the atmosphere, which, as far as I can discover, is entirely new, and will be found, I believe, to solve all the phenomena of vapour we are acquainted with." The theory was that "evaporation and the condensation of vapour are not the effects of chymical affinities, but that aqueous vapour always exists as a fluid sui generis, diffused among the rest of the aerial fluids . . . there is no need to suppose a chymical attraction in the case." [47]

---

[45] Now preserved in Manchester College, Oxford. Davis (1932), *Manchester College*, is a competent survey, but throws little light on the Academy's turbulent early years, before the 1803 move to York.
[46] Smith (1856), *Memoir of Dalton,* 18, and Roscoe (1895), *Dalton,* 51.
[47] Dalton (1793), *Meteorology,* vi and 132–136.

In denying the chemical attraction of water for air in which it was "dissolved," Dalton was of course flouting the view that short-range attractive and repulsive forces were the appropriate media for explaining the process of evaporation. In support of such a view, chemists could quote no less an authority than Lavoisier, as well as the whole weight of Newtonian tradition. Dalton, with his "habit of looking upon all empirical phenomena from a mathematical point of view," [48] was not the one to worry about this. His experiments seemed to show that the absorption of water vapor by air *is not pressure dependent,* i.e., "that a cubic foot of dry air, whatever its density be, will imbibe the same weight of vapour if the temperature be the same." Such a conclusion (in our terms "that the vapor pressure of water is constant at constant temperature") could not easily be reconciled with belief in evaporation as a chemical process—hence Dalton, the mathematically inclined meteorologist, simply abandoned the chemistry!

In the appendix to the work, he went even further, saying that the vapor not only of water but "probably of most other liquids" too, existed at all times in the atmosphere in an independent state.[49] The ideas that in a mixture of gases each component acts as an independent entity ("Dalton's law of partial pressures") and that the air is *not* a vast chemical solvent were thus first presented in the 1793 *Meteorological Observations.* The statements brought no immediate reaction. This was only to be expected, Dalton's arguments being tentative and undeveloped, his presentation lacking both vigor and the aid of clarifying diagrams, the ideas themselves being so curious in a world of all-pervasive chemical forces, and the author and vehicle of publication comparatively obscure. Things were different eight years later when John Dalton, author of important papers on color-blindness and heat, was systematically to restate his views in both the monthly scientific journals of the day.

[48] The remark comes from W. C. Henry, Dalton's pupil and first biographer: see Thackray (1967), "Dalton Letters," 147.

[49] Dalton (1793), *Meteorology,* 200–202.

Papers that Dalton read to the Manchester Literary and Philosophical Society in 1799 and 1800 show how much the question of water vapor continued to exercise him. The really dramatic development came in the summer of 1801. By September he was confident enough in his ideas to take the bold step of writing to the recently established *Journal of Natural Philosophy, Chemistry and the Arts*. This showed no hesitation in publishing the "New theory of the constitution of mixed aeriform fluids, and particularly of the atmosphere" at which Dalton had arrived.[50]

That Dalton was convinced of the value of his ideas is apparent. The rough sketch of his "theory of mixed gases" afforded in the letter to the journal was quickly supplemented in three papers to the Manchester society. These included the clear statement that "when two elastic fluids, denoted by A and B, are mixed together, there is no mutual repulsion amongst their particles; that is, the particles of A do not repel those of B, as they do one another. Consequently, the pressure or whole weight upon any one particle arises solely from those of its own kind"[51] (see Fig. X). The debt of this generalized "new theory" to Dalton's earlier thinking is obvious. In 1793 he had been content to insist that "when a particle of vapour exists between two particles of air let their equal and opposite pressures upon it be what they may, they cannot bring it nearer to another particle of vapour, without which no condensation can take place." Now in 1801 air and vapor—or A and B—were not only stated to have no mutual *effect,* but there was to be no actual repulsive force between them.

While Dalton's earlier statement passed unnoticed, the reaction to the new theory of mixed gases was rapid and widespread. The long and carefully argued papers he published in the *Manchester Memoirs* were widely abstracted and reprinted. Discussion was immediate and lively. C. L. Berthollet was then

---

[50] Dalton (1802b). See the fuller account in Thackray (1966c), "The emergence of Dalton's theory."

[51] Dalton (1802c), "Experimental essays," 536.

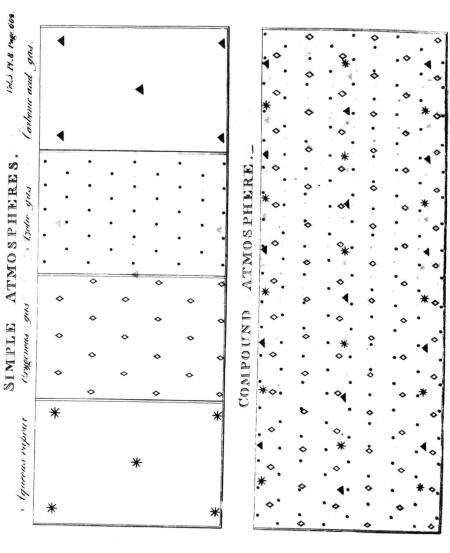

X. Dalton's illustration of his 1801 "theory of mixed gases." From Dalton (1802 c), 602.

busily at work on his affinity investigations, under Newtonian inspiration. He scornfully dismissed Dalton's diagramatic representation of the new theory as "un tableau d'imagination." Humphry Davy in turn felt constrained to write to a friend about "some papers of Mr. Dalton on the constitution of the atmosphere . . . executed in a very masterly way. I wish very much to have your judgment upon his opinions, some of which are new and very singular." Even the Literary and Philosophical Society was uncertain what to make of its secretary's dismissal of chemical affinity as a force acting in the atmosphere.[52] More damagingly, the 1802 first edition of Thomas Thomson's highly successful *System of Chemistry* was very critical of these new ideas. Dalton quickly wrote to both the monthly scientific journals of the day, rebutting Thomson's criticism,[53] but clearly it was not argument that was needed so much as convincing experimental proof of his beliefs. To provide such proof became Dalton's major aim, and the efficient cause of the chemical atomic theory. What began as a particular interest in meteorology thus ended up as a powerful and wide-ranging new approach to the whole of chemistry, though the transition was by no means sudden.

One thing Dalton did in order to provide support for his heavily attacked theory of mixed gases was to begin an experimental inquiry into the proportions of the various gases in the atmosphere. It was this inquiry which accidentally raised the whole question of the solubility of gases in water. By November 1802 he had progressed far enough to read to the Manchester society a paper "On the proportion of the several gases or elastic fluids, constituting the atmosphere; with an enquiry into the circumstances which distinguish the *chymical* and *mechanical* absorption of gases by liquids." When read, though not when published, this contained the statement that "[carbonic acid] gas is held in water, not by chemical affinity, but

[52] See Berthollet (1803), *Statique Chimique*, I, 499; Paris (1831), *Life of Davy*, I, 157; Henry (1804), "Mr. Dalton's theory," 29.
[53] Dalton (1802d), "Theory of mixed gases elucidated."

merely by the pressure of the gas . . . on the surface, forcing it into the pores of the water."[54] The accidentally begun researches on solubility thus led to an extension of his affinity-denying or "mechanical" ideas.

It seems that it was this extension of Dalton's ideas that provoked his close friend, the Edinburgh trained chemist William Henry, to begin his own rival and chemically orthodox "extensive series of experiments, with a view to ascertain the order of affinities of gases for water." Measured with reference to this objective, the experiments were not a success. However, within a month Henry had found what Dalton failed to see, namely "the following general law: that, under equal circumstances of temperature, water takes up, in all cases, the same volume of condensed gas as of gas under ordinary pressure." Aware of this work, and quick to see its relevance to his own ideas, Dalton was able to point out faults in Henry's procedure. One consequence was the latter's public admission that "the theory which Mr. Dalton has suggested to me on this subject, and which appears to be confirmed by my experiments, is that the absorption of gases by water is purely a mechanical effect."[55] ("Mechanical" here, as in Dalton's own usage, had a very specific meaning. To Dalton a "mechanical" process of absorption or evaporation was not something purely kinematic, as it had been to seventeenth-century corpuscular philosophers. Yet, though forces were involved, "mechanical" processes could be sharply differentiated from those depending on chemical affinity. The latter implied an interaction of forces, with a wholly new chemical—and force—result. In contrast a "mechanical" effect depended on selective, but unchangeable, repulsive forces of the sort suggested in Dalton's theory of mixed gases. No chemical change took place: no chemical forces were involved.)

In light of Henry's public support, we can appreciate why Dalton continued to grapple with "the absorption of gases by

[54] See Lonsdale (1874), *Dalton*, 310; and Dalton (1802d), 271.
[55] Henry (1803), "Solubility of gases," 41 and 274.

water." A further paper with that title was ready for the Manchester society in October 1803. In this paper he made it clear that though his theory of mixed gases was much strengthened by the new evidence from solubility studies, yet "the greatest difficulty attending the mechanical hypothesis arises from different gases observing different laws." Or, to put the problem in its crudest form: "Why does water not admit its bulk of every kind of gas alike?" It was to answer this question that Dalton developed what grew into the *New System of Chemical Philosophy*.

As he went on to report:

> This question I have duly considered, and though I am not yet able to satisfy myself completely, I am nearly persuaded that the circumstance depends upon the weight and number of the ultimate particles of the several gases; those whose particles are lightest and single being least absorbable and the other more according as they increase in weight and complexity.* An enquiry into the relative weights of the ultimate particles of bodies is a subject, as far as I know, entirely new; I have lately been prosecuting this enquiry with remarkable success. The principle cannot be entered upon in this paper; but I shall just subjoin the results, as far as they appear to be ascertained by my experiments.[56]

The paper—as printed in 1805—closed with a list of what we would now call "atomic weights," as well as with the disconsolate footnote that "*subsequent experience renders this conjecture less probable."

Dalton's method of calculating the "relative weights of ultimate particles" was simplicity itself. Unlike the Newtonian chemists—of Buffonian, Boscovichean, or no particular persuasion—the measurement or calculation of interparticle affinity

---

[56] Dalton (1805b), "Absorption of gases by water."

263

Atoms and Powers

forces held no interest for him. Instead his mechanistic, visual, and realist view of atoms was joined with the prevailing vogue for numerical calculation, and the common assumption of one to one combination, in such a way as to yield wholly new insights.

That Dalton quite literally believed in solid, spherical atoms combining in accord with simple mechanistic considerations is apparent from the diagrams with which he carefully furnished his *New System*. Later he even went so far as to have solid wooden "atoms" constructed, with spokes at appropriate angles, to illustrate his ideas (see Figs. XI and XII). The irony is that it was not until much later in the nineteenth century that chemists as a group were willing to accept the legitimacy of diagrams such as Dalton proposed. His realist insistence on the actuality of material atoms and their directional linkage diverted immediate attention from the very considerable chemical utility of three-dimensional models. (In the twentieth century, when chemists have far stronger experimental grounds for rejecting Dalton's solid atoms, models such as he first employed enjoy a new lease of life: see Fig. XIII.)

To Dalton himself, it was self-evident that *"binary* compounds must first be formed in the ordinary course of things, then *ternary,* and so on." The reason was not any arbitrary "rule of greatest simplicity" (as commentators have often assumed), but instead the quite straight forward consequence of Dalton's view of forces and chemical combination. Thus when A displays a chemical affinity for and unites with B there is "no mechanical reason" why one atom of A should not combine with "as many atoms of B as are presented to it, and can possibly come into contact with it." It is "the repulsion of the atoms of B among themselves" that actually limits the process. "Now this repulsion begins with 2 atoms of B to one of A, in which case the 2 atoms of B are diametrically opposed; it increases with 3 atoms of B to 1 of A, in which case the atoms of B are only 120° asunder; with 4 atoms of B it is still greater as the distance is

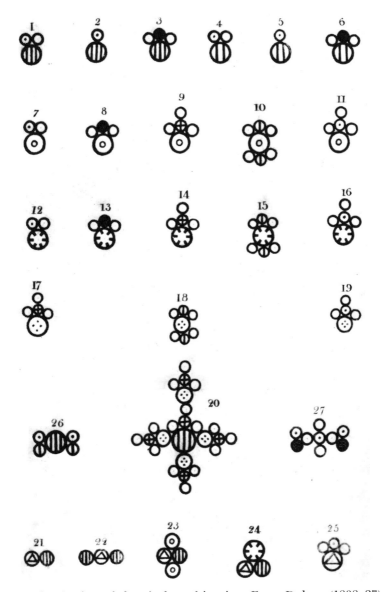

XI. Dalton's view of chemical combination. From Dalton (1808–27), plate VI.

then only 90°; and so in proportion to the number of atoms."[57] Obviously binary compounds were more likely to be formed than ternary, if Dalton's suppositions were correct. Armed with such a mechanical view of combining ratios, it was a simple matter to arrive at relative particle weights. If water was formed by the combination of oxygen with hydrogen in the ratio of eight ounces to one, then clearly the relative weights of their ultimate particles were as eight to one.

Just how little Dalton or anyone else realized the implications of his work is seen from public reaction to his tables. Though published in the *Manchester Memoirs,* and reprinted in both monthly scientific journals, his table of weights—unlike his theory of mixed gases—aroused no reaction at all. This shows how, by themselves, tables of weight numbers appeared to be just further obscure and unexplained variations on the widely known tables of affinity numbers. Even when accompanied with an explanation of their significance, a favorable reception was by no means certain. In December 1803, thanks to a slip in his arithmetic, Dalton was able to show Davy how the various oxides of nitrogen might be given formulae and particle weights that were in harmony with the latter's own experimental results. Yet Davy—true to his deeper Newtonian vision—simply

[57] Dalton (1811b), "Atomic principles of chemistry," 147. Dalton's "naive-realist" belief in three-dimensional atomic structures, and the unease of other chemists, is particularly apparent from *Rep. Brit. Assoc. Adv. Sc.* (1836) *4,* 207: "Dr Dalton stated to the chemical section his reasons for preferring the symbols which he had himself used from the commencement of the atomic theory in 1803 to the Berzelian system of notation subsequently introduced. In his opinion regard must be had to the arrangement and equilibrium of the atoms (especially elastic atoms) in every compound atom, as well as to their number and weights. A system of either *arrangements* without *weights,* or of *weights* without *arrangements,* he considered only half of what it should be." The less formal *Proceedings of the Fifth Meeting* (Dublin, 1835) provide an even more graphic picture (p. 82): "Dr Dalton next brought forward his views of the atomic theory, which appeared to excite great interest. . . Dr Dalton had prepared a lithographic plan of his arrangement, and showed how the molecules of bodies may be considered grouped so as to represent compound atoms. . . He stated that he considered this method as *the only one representing nature. . .* Professor Whewell stated, that reserving all due respect to the opinions of Dr Dalton . . . Dr Dalton's method supposes a theory . . . [and] attempts to show [the atoms'] method of molecular arrangement, of which we have no positive knowledge whatsoever." (Italics added.)

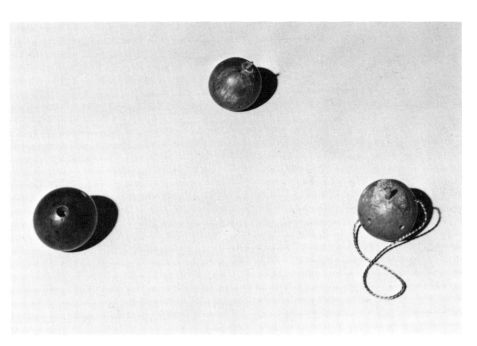

XII. Dalton's wooden atomic models. Now in the Science Museum, London. Photo: British Crown Copyright. Science Museum, London.

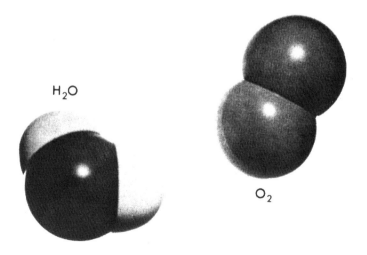

XIII. Modern atomic models. From PSSC PHYSICS, 2nd ed. Copyright ©, 1965, by Education Development Center. Published by D. C. Heath and Co., Boston, Mass.

dismissed Dalton's ideas as speculations "rather more ingenious than important." [58]

This lack of enthusiasm for the chemical possibilities of his work must have been a blow to Dalton—Davy was after all by far the most capable, serious, and ambitious chemist he had so far encountered. However it is not clear that Dalton himself had yet fully grasped the wider implications of his work. In 1804 he did succeed in arriving at formulae for various hydrocarbons which were agreeable both to his calculating system and his own—now rapidly increasing—chemical experiments. But 1804 was chiefly noticeable for continuing controversy over the mixed gases theory, and particularly over its denial of weak chemical affinity forces. Continuing criticism of the theory—and the failure of particle weight studies to provide the hoped-for clinching evidence—caused Dalton to revise his ideas on mixed gases. It was this revision which seems to have led to the slowly deepening conviction that his work on particle weights, though not a success in its original purpose, could well form the basis of a *New System of Chemical Philosophy*.

In the syllabuses of the public lecture courses he gave in London late in 1803, in Manchester early in 1805, and in Edinburgh in April 1807, we can trace the slow shifting of Dalton's interest away from mechanics, meteorology, and mixed gases, towards chemistry. By March 1807 he was writing to Thomas Thomson to offer an Edinburgh lecture course. This was to be on his "late experimental enquiries," including "chemical elements or atoms with their various combinations," and would reveal "my latest results, some of which have not yet been published or disclosed in any way, & which I conceive of considerable importance." The lectures duly took place. By this time Dalton's introduction left no doubt as to the importance of his ideas for chemistry:

> I have been enabled to reduce a number of apparently anomalous facts to general laws, and to exhibit a new view of the

---

[58] See Roscoe and Harden (1896), *New View of Dalton*, 44, and Thackray (1966b), "Documents on atomic theory," 3.

first principles or elements of bodies and their combinations, which, if established, as I doubt not it will in time, will produce the most important changes in the system of chemistry, and reduce the whole to a science of great simplicity, and intelligible to the meanest understanding.[59]

The lectures in Scotland were little short of a manifesto for the *New System*. That their reception was favorable we know from the dedication attached to the first part of that work when it finally appeared, just over a year later. This first part of a never completed enterprise must have struck its chemical readers as a curious production, revealing as it did the research activities over the previous eight years of a provincial meteorologist and natural philosopher. The work ran to 220 pages. The first 208 were occupied with discussion of heat, mixed gases, and gas-liquid solubility. Only the last thirteen pages were devoted to "Chemical synthesis." Dalton did just find time to state his views on chemistry, though his table of atomic weights was only appended to the last plate at the back of the book!

With the publication of the second part of the *New System* in 1810, and more especially with Thomson and Wollaston's 1808 papers showing the practical power of his approach,[60] we may consider Dalton's ideas as launched. It is beyond the scope of this work to examine the reaction of practicing chemists to those ideas, or their profound influence on the whole of nineteenth-century science. But in conclusion we must explore a little further what exactly Dalton's chemical atomism involved. This examination will reveal something of the confused nature of Dalton's own thought on the nature and properties of matter, and show the equivocal status held by chemical atoms right from the start of their nineteenth-century career. It will also help us to understand how Newtonian ideas lingered on in

---

[59] For the lecture syllabuses, see Thackray (1966b); for the letter to Thomson, Thackray (1967), "Dalton's letters," 162; for Dalton's introduction, Roscoe (1895), *Dalton,* 167.

[60] Thomson (1808), "On oxalic acid," and Wollaston (1808), "On super-acid and sub-acid salts."

spite of Dalton's work, even though such ideas were hostile to a strictly "Daltonian approach."

8.6. A New System of Chemical Philosophy?

It has been necessary to describe Dalton's background and the development of his ideas in some detail. We can now appreciate just how a "new system of chemical philosophy" originated from a man so unversed in chemistry and so remote from the issues of chemical debate. Dalton was manifestly no Boscovich possessed of striking and lucid powers of argument and persuasion, nor yet a Buffon intent on revealing the universal applicability of nature's fundamental law of force. Still less was he a Guyton or a Kirwan, in pursuit of a quantified Newtonian science of chemical mechanisms. Yet he twice appealed to Newton in support of his ideas,[61] as if Newton were the final, undisputed court of appeal. Ironically his appeal was in both cases founded on an untenable reading of Newton's text. How then should we view Dalton?

Perhaps most important is to recognize that Dalton's whole training and intellectual formation was in a context remote from chemical discussion and debate. His early theoretical investigations were in the field of applied mathematics, and his practical work in meteorology and natural history. The popular Newtonianism of textbook and lecture tradition provided his immediate reference frame, with common-sense philosophy as its ontological support. In addition he was aware of the unease over the implications of Newtonianism, common to the more cautious and devout Christian groups. Belief in peculiarly solid and realistic atoms of inert matter thus came naturally to him, as did an almost total antipathy to the investigation and discussion of chemical forces. He seems at best to have been indifferent to the sophistication of the orthodox Newtonian chemistry of his day. Conversely, when the moment came, he was not afraid to draw on and exploit that denial of inertial homo-

[61] See below, note 68.

geneity in matter which had lurked among the "theological anti-Newtonians" for much of the century. To these outside conditioning factors we must add his own gifts of great tenacity, a highly visual imagination, and a brilliant ability for devising appropriate experimental investigations. He was also never afraid to generalize, mathematize, and expound his ideas, and thus was fully willing to reorder chemistry when other pursuits led him to the task.

His first biographer was to remark (though not in public) on Dalton's "habit of looking upon all empirical phenomena from a mathematical point of view." Coupled with this was "his eminently self-reliant nature and a certain unbendingness of will, near of kin to obstinacy." [62] This "unbendingness of will" was crucial to the formation of an atomic theory driven through despite the recalcitrance of Nature. The "Pythagorean" love of mathematical relationships—apparent time and again in Dalton's work—was also to play its part. It is thus not surprising that though the "new system" he produced had an almost startling visual clarity, and undeniable explanatory power, its philosophical foundation was a matter of ambiguity and confusion.

8.6.1. Chemical Units

Since the equation of the words "atom" and "chemical element" is usually held to be one of the most important aspects of the *New System,* and was certainly one that led to continuing controversy, it is of some importance to inquire into Dalton's thinking on this subject. In the opening sections of the present essay, Newton's ideas and their philosophical basis in the unity of matter were explored at some length. The way in which these ideas in their "nut-shell" form came to underlie discussion of the nature, number, and stability of the chemical elements has also been displayed. The roots of Dalton's thinking would seem to lie in other areas. Yet to believe in solid, massy, inert atoms was one thing. To equate them on a one-to-one basis with the different chemical elements was something else again. It appears

[62] Thackray (1967), "Dalton letters," 146.

that Dalton's path to this latter position was slow and deliberate.

If we consult the syllabus for the course of lectures he gave at the Royal Institution in December 1803, soon after the first measurement of relative particle weights, we read in thorougly orthodox fashion of "properties of matter. Extension—impenetrability—divisibility—inertia—various species of attraction and repulsion. Motion—forces—composition of forces—collision. Pendulums." Yet by the spring of 1805, under pressure from the continuing success of his chemical investigations, Dalton's ideas were changing. The newly discovered syllabus for the lectures he gave in Manchester at that time begins: "Introduction.—General properties of matter—extension—divisibility—original ideas on the division of matter into elements and their composition—solidity—mobility—inertia." [63] By this time also, one whole lecture was "On the elements of bodies and their composition." Tantalizingly, no manuscript survives to enlighten us about the "original ideas on the division of matter into elements." However it seems reasonable to suppose that as Dalton slowly came to appreciate the far wider chemical utility of his researches on "the relative weights of ultimate particles," so he increasingly felt the need to define the nature of these ultimate particles. And thanks to his background and affiliations, the move to explicit avowal of chemical atoms and heterogeneous matter was a comparatively simple one to make.

It is in the 1807 Edinburgh syllabus that we find the first direct mention of "indivisible particles or atoms." Lectures 3, 4, and 5 of Dalton's now deliberately propagandizing course were "Of chemical elements." The syllabus began: "Elastic fluids conceived to consist of indivisible particles or atoms of matter, surrounded with atmospheres of heat.—Liquid and solid bodies conceived likewise to be composed of atoms surrounded with heat." [64] Even this statement was not as unambiguous as might be supposed. Of the eighteen elastic fluids known to Dalton, fif-

---

[63] Thackray (1966b), "Documents on atomic theory," 12, 15, and 16.
[64] *Ibid.*, 20.

teen were—by his own reckoning—compounds. Thus Dalton was in part using the word "atom" in the commonplace and acceptable sense of "smallest particle possessing a given nature." In this sense, atom was merely a term for a particle which was *divisible only with the loss of its distinguishing chemical characteristics*. Yet Dalton's position was not so clear cut. He was also beginning not only to think but to speak in public of chemical atoms in the more radical sense of *solid and indivisible* particles.

Part I of the *New System* was to say that "chemical analysis and synthesis go no further than to the separation of particles one from another, and to their reunion. No new creation or destruction of matter is within the reach of chemical agency." So far this was a statement unlikely to raise objection from Newton, Stahl, Lavoisier, or any competent chemist. What was new was the further insistence that "we might as well attempt to introduce a new planet into the solar system, or to annihilate one already in existence, as to create or destroy a particle of hydrogen. All the changes we can produce, consist in separating particles that are in a state of cohesion or combination, and joining those that were previously at a distance." [65] In this way Dalton first made formal claim for the privileged status of his chemical atoms. The particle of hydrogen was not to be seen as the complex result of an ordered and intricate internal structure, but as the "given" solid, the planet.

Just why Dalton should have moved to this position of claiming privileged status for his chemical atoms is not fully obvious. No doubt he felt the need of some "philosophical" justification for his concentration on particle weights at a time when the list of elements was under renewed—electrochemical—attack. After all, Lavoisier's reforms had by no means settled the question of which substances should be admitted to the status of chemical element. Nicholson's 1795 *Dictionary of Chemistry* accurately reflected the prevailing mood when it said that "at present we hear little concerning elements. Those substances

[65] Dalton (1808–27), I, pt. I, 212.

which we have not hitherto been able to analyse ... are indeed considered as simple substances relative to the present state of our knowledge, but in no other respect." [66] Between 1800 and 1812 no less than fifteen new chemicals were added to the list of eighteen previously known elements. We can therefore appreciate how widely acceptable was Davy's impatient belief that whereas the power of Nature was limited "the powers of our artificial instruments seem capable of indefinite increase," so that "there is no reason to suppose that any real indestructible principle has yet been discovered." [67]

Having adopted a position, Dalton was not the one to settle for half measures. In his 1810 lectures at the Royal Institution —lectures clearly designed to defend and vindicate his now widely known and controversial ideas against Davy and a host of critics—he publicly abandoned the unity of matter. Of particular interest is the way his views parallel the earlier expressions of George Adams, even to the "reinterpretation" of Newton to fit the new position. Dalton was prepared to admit how "it has been imagined by some philosophers that all matter, however unlike, is probably the same thing." However, on the excellent principle that attack is the best form of defense, he calmly asserted that "this does not appear to have been [Newton's] idea. Neither is it mine. I should apprehend there are a considerable number of what may be called *elementary* principles, which never can be metamorphosed, one into another, by any power we can control."

Still on the offensive, Dalton reiterated the same beliefs in print in 1811. He insisted that "atoms of different bodies may be made of matter of different densities." The example he offered was that "mercury, the atom of which weighs almost 170 times as much as that of hidrogen, I should conjecture was

---

[66] See Nicholson (1795), I, 155.

[67] Davy (1839–40), *Works*, V, 54, and IV, 132. It may appear paradoxical that an increase in the number of chemical elements was held to show that no "real indestructible principle" had been discovered. However there is a striking present-day analogy in the way nuclear physicists take the multiplying discoveries of elementary particles to show that these particles cannot "really" be elementary!

larger, but by no means in the proportion of the weights, which would require a diameter of five or six times the magnitude." Once again, the opposition was disarmed with the bland assertion that "perhaps in a question of this sort Newton has a better claim to be heard" than either Dalton or his critics. And, of course, "[Newton] says (I think in the 31st query to his *Optics*) 'God is able to create particles of matter of several sizes and figures, and in several proportions to the space they occupy, and perhaps of different densities and forces . . . at least I see nothing of contradiction in all this.' "[68]

The interesting thing about this last quotation, is what Dalton chose to omit. Newton was far from advocating the heterogeneity of the matter in our own world, as the missing words make plain. These missing words are "and thereby to vary the laws of Nature, and make worlds of several sorts in several parts of the universe." But Dalton was obviously concerned to utilize Newton in his own defense, not to quote him accurately. It was polemically useful to cite the *Opticks* in favor of elementary principles which could not be metamorphosed and—in corollary—atoms of different bodies made of matter of different densities. However the roots of such thinking would seem to lie rather in such writers as William Jones and George Adams than in Newton. Explicitly to avow this was scarcely possible. Hence the smoke screen round Dalton's ideas. Hence also the resulting confusion, and the unwillingness of so many chemists fully to embrace chemical atoms, the utility of which they appreciated, but the ontological base of which they could not understand.

Anyway, Dalton did defend his "chemical atoms," if only in a cursory way and at the price of the abandonment of both the unity of matter, and the Newtonian tradition of belief in an internal structure to the elements. Dalton's provision of a conceptual basis to chemistry independent of—indeed hostile to—the continuing beliefs of the physicists, was to prove enormously

---

[68] Dalton's 1810 statement is reproduced from the manuscript, as transcribed in Roscoe and Harden (1896), *New View of Dalton*, 112. See also Dalton (1811b), 150. On both occasions, the same misquotation of Newton was employed. Dalton's remarks should also be compared with those of William Jones: see note 33 above.

valuable throughout the nineteenth century. His obliteration of the divide that separated ultimate atoms from chemical elements was to open a new era. The visual utility of his work was to prove enormous, especially later in the century as organic chemistry developed. Indeed the triumphs of the chemical atomic theory were so great as to hide from many later chemists the ambiguous nature of Dalton's own writings and the uncertainty of his defense of chemical atoms. In light of our study of eighteenth-century thought, we can understand both the mixed reception that Dalton's ideas received, and the continuance of a stream of "anti-Daltonian" speculation throughout the nineteenth century.

8.6.2. Chemical Mechanisms

Dalton's quantification of chemical units was operationally successful, if philosophically awkward. His approach to the question of chemical mechanism was far more negative. As far back as 1793 his ideas on the atmosphere had been based on a purely "mechanical" approach, with a *denial* of the existence of any sort of chemical affinity forces. The 1801 "theory of mixed gases" had been even more extreme, postulating a separate repulsive power peculiar to each individual gas, and making no attempt at all to explain why chemical reaction should ever occur. His 1805 revision of the theory [69] relied solely on the repulsive power of heat, and again failed to explain why chemical reaction ever happened. So too with the *New System* itself. Small wonder that such chemists as Davy and Berthollet, preoccupied with the problems of chemical mechanism, should find Dalton's ideas hard to credit.

Indeed in November 1806, at the very period when Dalton was formulating his system, Davy—at the peak of his Voltaic triumphs—was informing a fascinated Royal Society that the relation of electrical energy "to chemical affinity is . . . sufficiently evident. May it not be identical with it, and an essential property of matter?" If this were admitted, then Berthollet's

[69] See Thackray (1966c), "The emergence of Dalton's theory."

work on the influence of mass could be reconciled with older ideas on affinities. Indeed "Guyton de Morveau's experiments on cohesion" could even be reinterpreted to agree with the new electrical studies! [70]

In contrast to such speculations, the great utility of Dalton's method of quantification was widely and quickly appreciated. As early as January 1808, when only Thomas Thomson's short account of Dalton's system was available, W. H. Wollaston was declaring that "the inquiry which I had designed appears to be superfluous, as all the facts that I have observed are but particular instances of the more general observation of Mr. Dalton." And with all the zeal of a partisan, Thomson was to claim in 1825 that "several chemical manufacturers in this country have already availed themselves of the atomic theory to rectify their processes. This they have done to such an extent, that unless all the other manufacturers follow their example, they will find it impossible to compete with their more skilful rivals." [71]

The obvious success of Dalton's work on the one hand, and the awkward implications of Berthollet's critique on the other, were to signal the end of the century-long Newtonian vision of a quantified science of chemical mechanisms. Dalton's work was to shift the whole area of philosophic debate among chemists from that of chemical *mechanisms* (the *why?* of reaction) to that of chemical *units* (the *what?* and *how much?*). The replacement of the fruitless endeavor to quantify the forces of chemical *mechanism* by his astonishingly successful weight-quantification of chemical *units*, undercut the whole Newtonian program. Only with hindsight is it clear that Dalton's was the more fruitful path to tread, and chemists of his own day had every reason to complain about his brusque abandonment of mechanism studies. The assessment which Davy made of his rival's merits, when presenting him with the first Royal Medal of the Royal Society, well illustrates the reaction of the majority of chemists.

[70] Davy (1839–40), *Works*, V, 39–41 and 51.
[71] Wollaston (1808), "On super-acid and sub-acid salts," 96; Thomson (1825), *First Principles of Chemistry*, I, iii.

After mentioning "those Newtonian philosophers who supported the permanency of atoms, and their uniform combinations, such as Keill [and] Freind," Davy elaborated on the merits of Bryan and William Higgins. He then went on: "Mr. Dalton's permanent reputation will rest upon his having discovered a simple principle, universally applicable to the facts of chemistry—in fixing the proportions in which bodies combine, and thus laying the foundation for future labours, respecting the sublime and transcendental parts of the science of corpuscular motion. His merits in this respect resemble those of Kepler in astronomy."

This might appear high praise. Yet the analogy with Kepler was deliberate. True to his youthful beliefs, Davy was still convinced of the primacy of mechanism studies. As he made quite plain, "the causes of chemical change are as yet unknown, and the laws by which they are governed." His own belief was that in the developing "connection with electrical and magnetic phenomena, there is a gleam of light pointing to a new dawn in science." Even so the day still lay ahead when chemistry would "find some happy genius, similar in intellectual powers to the highest and immortal ornament of this Society [Newton], capable of unfolding its wonderful and mysterious laws." [72]

Such was the value of Dalton's stress on billiard ball atoms and combining weights that within less than a generation Edward Turner [73] could claim that "atomic theory ... contains the leading principles of the science, and forms the distinguishing feature, as it is the pride, of modern chemistry." That chemistry needed a new Newton was also not undisputed. Charles Daubeny [74] dedicated his 1831 *Introduction to the Atomic Theory* to "John Dalton ... the author of a theory with respect to the mode of combination between bodies, which stands foremost among the discoveries of the present age, for the universality of its applications, and the importance of its practical results; hold-

---

[72] Davy (1839–40), *Works*, VII, 95 and 97–98.
[73] (1796–1837). See *Partington* (1961–64), IV, 227–228. The judgment on chemical atomic theory is from Turner (1825), *Introduction to Atomic Theory*, iii.
[74] (1795–1867). See *DNB*.

ing the same kind of relation to the science of chemistry, which the Newtonian system does to mechanics."

Such early enthusiasm by minor British chemists was in due course reinforced by the verdict of weightier judges. By 1853 no lesser scientist than Justus Liebig was to pronounce that

> All our ideas are so interwoven with the Daltonian theory, that we cannot transpose ourselves into the time, when it did not exist ... Chemistry received in the atomic theory, a fundamental view; which overruled and governed all other theoretical views... In this lies the extraordinary service which this theory rendered to science, viz.: that it supplied a fertile soil for further advancements; a soil which was previously wanting. In the most recent investigations concerning the constitution of organic bases, the alcohols and the acids corresponding to the alcohols, we have seen that the groundwork of the Daltonian theory is equally valid for organic bodies.[75]

The majority of those responsible for the chemical triumphs of the half-century following Liebig's pronouncement were to concur in his judgment of Dalton's theory as at once "a fundamental view" and "a fertile soil for further advancements." Indeed the continuing success of the theory was to blunt the thrust of all nineteenth-century attempts to criticize its fundamental assumptions.[76] To this extent the *New System of Chemical Philosophy* deserved its title. Dalton's stress on macro-scale weights and his contempt for mechanism studies both spelled the end of a century-long Newtonian tradition in chemistry, and pointed a fruitful new way forward.

[75] Quoted from Henry (1854), *Life and Researches of Dalton*, 134.
[76] See Buchdahl (1959), "Scepticism in atomic theory." The particular case of nineteenth-century English criticism is explored in Knight (1967), *Atoms and Elements*.

9
Conclusion

> Whatever may be the ultimate fate of the theory which found deliberate expression in the *New System of Chemical Philosophy*—and no one can say that it is not destined to give place to a higher and even bolder generalisation, which shall more clearly connect matter with the forces associated with it—it is certain that the ages to come will reckon it as the central, dominant conception which has activated the chemistry of the nineteenth century. The characteristic feature of the chemistry of our time is, in a word, the development and elaboration of Dalton's doctrine; for every great advance in chemical knowledge during the last ninety years finds its interpretation in his theory.
> T. E. Thorpe in 1902

The successes achieved by chemical research in the nineteenth century form a striking contrast to those of the previous hundred years. The chemical atomic theory was fundamental to these successes. Whatever the private and continuing doubts of many practitioners about its ultimate validity, this theory did provide both a secure ontological basis for the science and an astonishingly successful model for chemists to use in their researches. The theory was profoundly antiphysicalist and anti-Newtonian in its rejection of the unity of matter, and its dismissal of short-range forces. The triumphs of nineteenth-century chemistry were thus built on no reductionist foundation, but rather achieved in isolation from the newly emerging science of physics. That this divorce could not last is immaterial. What is important is that chemical science in a fully useful form progressed on the basis of its own peculiar model—the chemical atom—and its own particular technique—that of macro-scale weight studies, not micro-scale concern with forces.

This conclusion is of some significance. It exposes but one of the many problems associated with those fashionable approaches that would see the growth of modern science as uniquely linked to the seventeenth-century triumphs of corpuscular philosophy, and its associated "mechanization of the world picture." The ambiguous and unsatisfactory relationship between chemistry and Newtonianism which this present essay has been concerned to explore, reveals something of the weakness of such reductionist approaches to the history of science. To see the achievements of Lavoisier and Dalton as central to a physicalist "Postponed Scientific Revolution in Chemistry" is to reveal a confusion of thought, and a distressing ignorance of the far more subtle routes by which chemistry developed. To understand these routes will require a greater attention than has so far been paid to Stahlian thought, natural history, Condillacean and common-sense philosophy, and anti-Newtonianism in all its variant forms.

On another level, the continuing and fruitful interaction between theology and the theory of matter presents problems to those who would see post-seventeenth-century science as an intellectually autonomous entity. At least in Britain, theological and scientific thinking about the nature and properties of matter were in a state of creative tension throughout the eighteenth century. The line of theological development from Newton through Priestley was, in its implications for any theory of matter, to reach at least as far into the nineteenth century as Michael Faraday. Conversely, the religious and anti-Newtonian speculations of the Hutchinsonians were to enjoy an enduring scientific impact thanks to their influence on the formulation of Dalton's chemical atomism. Again, the very different characteristics that Newtonian matter-theory assumed in France cannot be understood apart from the profound differences in religious context between the two countries.

Similar problems are posed by the definite impact of political controversy on the course of scientific research and debate. In this study we have seen how the Newton-Leibniz clash polarized natural philosophers throughout Europe into hostile groups. In this way a personal and national quarrel profoundly altered the

Conclusion

context of natural philosophy. But more than that, the episode had a *constitutive* effect on the course of research and debate. We have tried to show this in connection with the elaboration and acceptance of the characteristically Newtonian "nut-shell" view of matter. It may also be dimly discerned—if not yet investigated or demonstrated—in the failure of Newtonianism to capture any significant group of German chemists. The same constitutive effect can be seen in British reaction to the French Revolution, and with it to the "excesses" of Priestley's theory of matter. Rather different questions about the influence of politics are illustrated by the role of Buffon. His combination of powerful patronage and particular private enthusiasms is itself an important explanatory factor in any discussion of French Newtonianism and chemical theory. The influence of other groups, from Newton's early circle to Dalton's Quaker friends, invites attention.

In an entirely different way, the role of technology demands inclusion in any historical discussion of chemistry. Its influence on the course of research is often subtle and not easy to evaluate. In the present study we have noted one of its effects in the creation of a demand for popular lecture courses. The pressures placed on the theoretical Newtonianism of such figures as Peter Shaw and William Cullen by their roles as chemical technologists suggest some possibilities for further inquiry. The particular industrial and utilitarian context within which John Dalton began his self-education in chemistry also invites what may well be rewarding questions. And Thomas Thomson's claim of how quickly chemical manufacturers availed themselves of the atomic theory suggests still further lines of research into the relationship between scientific theory and industrial activity.

The influence of theological considerations was especially critical to the development of matter-theory at particular moments. So too the pressure of political events was far from uniform when it came to shaping the developing course of Newtonian thinking. And, from yet another angle, the technological concerns bearing down on all chemists were far different by the end of the century than they had been in 1700. Adequately and historically to illustrate the way these very diverse factors operated to the common end of shaping the development of Newtonian matter-theory is a task of some complexity. This essay has attempted to illuminate some of the most critical areas. The approach employed has necessarily been somewhat eclectic, yet not more so than the subtly shifting forces of history demand.

At this particular stage of historical inquiry, conjecture is far simpler than demonstration. We shall therefore conclude this essay by endorsing Newton's verdict that "there are things that cannot be explained in few words, nor are we furnished with that sufficiency of experiments which is required to an accurate determination and demonstration." But perhaps enough has been said to suggest that the development of chemistry in the early modern period was more complex than most accounts allow. A fuller grasp of this complexity will help transform our still too simple picture of the causes of scientific change.

Bibliographic Note
Select Bibliography
Index

Bibliographic Note

Possessing a range far wider than its short title indicates, Cohen (1956), *Franklin and Newton* stands as the fundamental biographical and bibliographical reference work for students of Newtonianism through the eighteenth century. The chemical writings relevant to this present essay are best approached via the rich chunks of information and comment packed into volume III of Partington (1961–64), *History of Chemistry*. The science of the period is surveyed in Wolf (1939), *Science, Technology, Philosophy in the Eighteenth Century*. English theological thought was long ago subject to the masterly analysis of Stephen (1876), *Eighteenth-Century English Thought;* Lange (1877–81), *History of Materialism* is the complementary source for Continental developments. The political context may best be approached through the relevant volumes of the *New Cambridge Modern History* as they appear, and through the *Oxford History of England*. For technological developments, the appropriate chapters of Kranzberg and Pursell (1967), *Technology in Western Civilization,* are of more immediate value than the encyclopedic treatment available in volume IV of Singer, Holmyard, Hall, and Williams (1954–58), *History of Technology*. In either case, Ashton (1948), *The Industrial Revolution,* is a classic that should also be consulted.

Needless to say, none of these works attempt that relating of Newtonian ideas on matter to the development of chemistry which has been the concern of this essay. Such a correlation is also foreign to the various surveys of the history of atomism, e.g., Gregory (1931), *Short History of Atomism,* and Van Melsen (1952) *Atomos to Atom*. Specialist monographs on atomic doctrines are either confined to

Bibliographic Note

the period before Newton or to chemical atomism after Dalton. The former are represented by Lasswitz (1890), *Geschichte der Atomistik,* and now also by Kargon (1966), *Atomism in England.* The latter include several recent studies: Brock (1967), *Atomic Debates,* and Knight (1967), *Atoms and Elements,* are each restricted to developments in England, but Farrar (1965), "Nineteenth-century speculations," is of European scope, as are such older studies as the appropriate sections of Freund (1904), *Chemical Composition,* and Muir (1906), *Chemical Theories and Laws.*

In contrast, theories of matter in the critical period from Newton to Dalton have been little studied. The shining exception that proves the rule is Metzger (1935), *La Matière chez Lavoisier.* The remarks in that short study are also illuminated by various comments in Metzger (1930), *Newton, Stahl, Boerhaave,* and Metzger (1938), *Attraction Universelle.* The present essay is, I believe, the first attempt ever made systematically to relate Newtonian views on matter to the development of chemistry, or to their wider background.

The pace of scholarly inquiry in the history of science is such that a variety of relevant studies have either appeared or been promised since my own research for this volume was concluded. Thus Professor I. B. Cohen has in press a long-awaited and sorely needed critical edition of the *Principia,* while J. E. McGuire continues to produce important articles concerning the evolution of Newton's thought on matter and its properties. Professor R. E. Schofield is also busy on a study of Newtonianism in the eighteenth century, which cannot fail to possess great relevance to this present inquiry; likewise Professor H. E. Guerlac's continuing work on Newton and Lavoisier. Finally I might note that the essays now available in Cardwell (1968), *Dalton and Science,* augment and complement my own investigations. Of the still other contributions which are even now (December 1968) being printed or planned, I am mercifully unaware.

Select Bibliography

All items referred to in the notes to the text are listed below. A number of primary and secondary sources to which no direct reference has been made are also included. These items are given a place because of their particular interest, relevance, or importance. In the case of Newton it has been impossible to include more than a fraction of the recent literature. Deliberate omission and unconscious neglect combine to ensure that coverage of other areas is far from exhaustive. Nonetheless I hope that this bibliography will be of use not only to historians of chemistry, but also to those concerned with other aspects of natural philosophy.

A. Manuscripts consulted

Joseph Black. Student's lecture notes dated 1767: Manchester University Library, Special Collections, Ms. CH B 106.

British Museum, Sloane Mss. A particularly rich source for the study of early-eighteenth-century natural philosophy. Includes letters by George Cheyne, J. T. Desaguliers, John Freind, David Gregory, Stephen Hales, John and James Keill, Richard Mead, Isaac Newton, Archibald Pitcairne, Peter Shaw, and Brook Taylor.

Cambridge University Library, Lucasian Mss. (uncatalogued). Include important manuscripts of J. T. Desaguliers, John Keill, and Henry Pemberton.

Select Bibliography

Christ Church, Oxford. Library. Mss. of David Gregory including his diary (mostly published in Hiscock [1937], *Newton's Circle*), material relating to J. T. Desaguliers, letters by John Freind, and some Newton items.

William Cullen. Student's lecture notes dated 1762–63: Manchester University Library, Special Collections, Ms. CH C 121 1–4. Student's lecture-notes dated 1757: Wellcome Historical Medical Library, London, Ms. 4674–75.

John Dalton. Dalton family Mss.: Cumberland and Westmorland Record Office, The Castle, Carlisle, Dalton (Eaglesfield) Mss., Ref. D/Da. Miscellaneous Dalton Mss.: The Strong Room, Friends' Meeting House, Kendal, Westmorland. Further Dalton Mss. as listed in Smyth (1966), *Bibliography*, and Thackray (1967), "Dalton letters."

John Hadley. Lecture notes: Trinity College Library, Cambridge, Ms. R I 50–51.

John Mickleburgh. Lecture notes: Caius College Library, Cambridge, Ms. 619/342 red.

Isaac Newton. Mss. in Cambridge University Library and King's College Library, Cambridge. Mss. by a variety of early-eighteenth-century figures are included: see Portsmouth (1888) and Munby (1949) for details.

Archibald Pitcairne. National Library of Scotland, Mss. 3440 ff. 20–27, and 3582 f. 66. Edinburgh University Library, Mss. Dc 1 61 and 62.

Royal Institution, London. Library. The *Managers' Minutes*, together with the holdings of Davy letters, throw light on English science in the first decade of the nineteenth century.

Royal Society of London. Library. Among their rich holdings, the volumes of the *Journal Book of the Royal Society*, and the corresponding *Council Book*, are particularly valuable sources. Also manuscripts by, to, or concerning, almost every active natural philosopher in the English-speaking world of the eighteenth century.

Edgar Fahs Smith Memorial Collection in the History of Chemistry, University of Pennsylvania. Important collection of letters including items by Joseph Black, John Dalton, Humphry Davy, Antoine Lavoisier, Joseph Priestley *et al.*

Wellcome Historical Medical Library, London. Rich and eclectic holdings of correspondence by or to a variety of eighteenth-century natural philosophers closely (e.g., Black) or remotely (e.g., Dalton) connected with medicine.

Select Bibliography

B. *Printed works* (An asterisk denotes works I have not seen)

Achard, F. C. (1776). Mémoire sur la force avec laquelle les corps solides adhérent aux fluides, où l'on détermine les loix auxquelles cette force est soumis . . . *Mém. Acad. Berlin*, 149–159.

Adams, G. (1794). *Lectures on natural and experimental philosophy . . . Describing . . . the principal phenomena of Nature, and shewing that they all cooperate in displaying the Goodness, Wisdom, and Power of God.* 5 vols. London.

Alexander, H. G. (1956). *The Leibniz-Clarke correspondence together with extracts from Newton's* Principia *and* Opticks, *edited with introduction and notes* . . . Manchester.

Ashton, T. S. (1948). *The industrial revolution. 1760–1830.* London.

Babbage, C. (1830). *Reflections on the decline of science in England.* London.

[Babson] (1950). *A descriptive catalogue of the Grace K. Babson collection of the works of Sir Isaac Newton.* New York.

———(1955). *A supplement to the catalogue . . . compiled by H. P. Macomber.* Babson Institute, Massachusetts.

Baron, T. (1756). Ed. *Cours de chymie.* Paris (New edition of Lemery [1675]).

Baumé, A. (1763). *Manuel de chymie* . . . Paris.

———(1778). *A manual of chemistry . . . translated from the French* [*by J. Aikin*]. Warrington.

Becker, C. L. (1932). *The heavenly city of the eighteenth-century philosophers.* New Haven, Conn.

Beer, C. de (1952). The relations between Fellows of the Royal Society and French men of science when France and Britain were at war. *Notes Rec. R. Soc. Lond.* 9, 244–299.

Bergman, T. (1775). Disquitio de attractionibus electivis. *Nova Acta R. Soc. Scient. Upsal.* 2, 159–248

———(1779–83) . . . *Opuscula physica et chemica* . . . 3 vols. Uppsala.

———(1788–90) . . . *Opuscula physica et chemica* . . . 6 vols. Leipzig.

———(1785). *A dissertation on elective attractions . . . translated from the Latin by* . . . [*T. Beddoes*]. London.

Berthollet, C. L. (1785). Observations sur l'eau régale et sur quelques affinités de l'acid marin. *Mém. Acad. R. Sci.*, 296–315 (published 1788).

Select Bibliography

———(1801).* *Recherches sur les lois de l'affinité.* Paris.

———(1803). *Essai de statique chimique* ... 2 vols. Paris.

———(1804). *An essay on chemical statics ... translated by B. Lambert.* 2 vols. London.

———(1809). *Researches into the laws of chemical affinity ... translated from the French by M. Farrell.* Baltimore, Md.

Besterman, T. (1953–65). Ed. *Voltaire's correspondence.* 107 vols. Geneva.

Black, J. (1803). *Lectures on the elements of chemistry ... Edited by J. Robison.* 2 vols. Edinburgh.

———(1963). *Experiments upon magnesia alba, quick-lime and other alcaline substances* ... Alembic Club Reprints, no. 1. Reissue edition. Edinburgh. (Originally published in 1756, in *Essays and observations, physical and literary. Read before a Society in Edinburgh and published by them.* 3 vols. [Edinburgh, 1754–71], II, 157–225.)

Bloch, E. (1913). Die antike Atomistik in der neueren Geschichte der Chemie. *Isis I,* 377–415.

Boas, M. See [Hall], M. B.

Boerhaave, H. (1719). *A method of studying physick. Containing what a physician ought to know in relation to the nature of bodies, the laws of motion ... chymistry, pharmacy and botany ... with the names and characters of the most excellent authors on all these subjects ... translated into English by Mr. Samber.* London.

———(1727). *A new method of chemistry; including the theory and practise of that art; laid down on mechanical principles, and accommodated to the uses of life ... translated from the printed edition, collated with the best manuscript copies. By P. Shaw, M.D. and E. Chambers, Gent.* London.

———(1732) *Elementa chemiae* ... 2 vols. Leyden.

———(1735). *Elements of chemistry: being the annual lectures of Herman Boerhaave ... translated from the original Latin, by Timothy Dallow, M.D.* 2 vols. London.

———(1740). *A treatise on the powers of medicines ... translated ... by John Martyn.* London.

Bonno, G. (1939). Deux lettres inédites de Fontenelle à Newton. *Mod. Lang. Notes 54,* 188–190.

Boscovich, R. J. (1763). *Theoria philosophiae naturalis redacta ad unicam legem virium in natura existentiam* ... Venice. (Original ed. Vienna, 1758.)

———(1922). *A theory of natural philosophy ... translated by J. M. Child.* London. (Facsimile reprint, London, 1966.)

Select Bibliography

Bouchard, G. (1938). *Guyton-Morveau. Chimiste et Conventionnel.* Paris.

Boyle, R. (1744). *The works of the Honourable Robert Boyle . . . to which is prefixed the life of the author.* Edited by T. Birch. 5 vols. London.
  See also Shaw, P. (1725).

Brewster, D. (1855). *Memoirs of the life, writings, and discoveries of Sir Isaac Newton.* 2 vols. London.

Brock, W. (1967). Ed. *The Atomic Debates. Brodie and the rejection of atomic theory. Three studies.* Leicester.

Brockbank, E. M. (1929). *John Dalton. Experimental physiologist and would-be physician.* Manchester.

———(1944). *John Dalton. Some unpublished letters of personal and scientific interest . . .* Manchester.

Brunet, P. (1926). *Les physiciens hollandais et la méthode expérimentale en France au XVIIIe siècle.* Paris.

———(1929a). *Maupertuis. Etude biographique.* Paris.

———(1929b). *Maupertuis. L'œuvre et sa place dans la pensée scientifique et philosophique du XVIIIe siècle.* Paris.

———(1931). *L'introduction des théories de Newton en France au XVIIIe siècle. Avant 1738.* Paris.

———(1936). Buffon, mathematicien et disciple de Newton. *Mém. Acad. Sci. Arts Dijon* 8, 85–91.

———(1952). *La vie et l'œuvre de Clairaut.* Paris.

Buchdahl, G. B. (1959). Sources of scepticism in atomic theory. *Br. Jnl. Phil. Sci.* 10, 120–134.

Bucquet, J. B. M. (1771). *Introduction à l'étude des corps naturels, tirés due règne minéral.* 2 vols. Paris.

Buffon, H. N. de (1860). *Correspondance inédite de Buffon, à laquelle ont été réunies les lettres publiées jusqu'à ce jour, recueillie et annotée . . .* 2 vols. Paris.

———(1863). *Buffon. Sa famille, ses collaborateurs, et ses familiers. Mémoires . . .* Paris.

Buffon, Comte de (1735). *La statique des végétaux et l'analyse de l'air . . . Par M. Hales . . . Ouvrage traduit de l'Anglois, par M. de Buffon.* Paris.

———(1740). *La méthode des fluxions, et des suites infinies. Par M. le Chevalier Newton* [translated by Buffon]. Paris.

———(1743). Dissertation sur les couleurs accidentelles. *Mém. Acad. R. Sci.,* 147–158 (published 1746).

———(1745a). Réflexions sur la loi de l'attraction. *Mém. Acad. R. Sci.,* 493–500 (published 1749).

———(1745b). Addition au mémoire qui a pour titre: réflexions sur

Select Bibliography

    la loi de l'attraction. *Mém. Acad. R. Sci.*, 551–552 (published 1749).

―――(1749–67). *Histoire naturelle, générale et particulière* . . . 15 vols. Paris.

―――(1775). *Histoire naturelle . . . Supplément.* Tome Première. Paris.

―――(1783–88). *Histoire naturelle des minéraux.* 5 vols. Paris.

―――(1785). *Natural history . . . translated by . . . W. Smellie.* 2nd ed. 9 vols. London.

Burton, W. (1746). *An account of the life and writings of Herman Boerhaave.* 2nd ed. 2 vols. London.

Burtt, E. A. (1925). *The metaphysical foundations of modern physical science. A historical and critical essay.* London.

Butterfield, H. (1949). *The origins of modern science.* London.

Cardwell, D. S. L. Ed. (1968). *John Dalton and the progress of science.* Manchester.

Carlid, G., and J. Nordström (1965). *Torbern Bergman's foreign correspondence.* Vol. 1. Stockholm.

Cassirer, E. (1951). *The philosophy of the Enlightenment.* Translated by F. C. A. Koelin and J. P. Pettegrove. Princeton, N. J.

Cavendish, H. (1921). *The scientific papers of the Honourable Henry Cavendish . . . edited from the published papers, and the Cavendish manuscripts.* 2 vols. Cambridge. (Vol. 1 edited by J. Clerk Maxwell, vol. 2 edited by E. Thorpe.)

Chaptal, J. A. C. (1790). *Elémens de chimie.* 3 vols. Paris.

―――(1791). *Elements of chemistry . . . translated by W. Nicholson.* 3 vols. London.

Châtelet, Madame la marquise du (1740). *Institutions de physique.* Paris.

Cheyne, G. (1702). *An essay concerning the improvements of the theory of medicine.* (Prefixed to the second edition of *A new theory of acute and slow continu'd fevers mechanically explain'd.*) London.

―――(1705). *Philosophical principles of natural religion: containing the elements of natural philosophy, and the proofs for natural religion, arising from them.* London.

―――(1715). *Philosophical principles of natural religion . . .* 2nd ed. London.

Cigna, G. F. (1772). Dissertation sur les diverses élévations du mercure dans les baromètres de différens diamètres . . . *Obs. Phys., Introduction 2,* 462–473.

Clagett, M. (1959). Ed. *Critical problems in the history of science.* Madison, Wisconsin.

Select Bibliography

Clairaut, A. C. (1739). Sur les explications cartesienne et newtonienne de la réfraction de la lumière. *Mém. Acad. R. Sci.*, 259–275 (published 1741).

——— (1745a). Du système du monde dans les principes de la gravitation universelle. *Mém. Acad. R. Sci.*, 329–364 (published 1749).

——— (1745b). Réponse aux réflexions de M. de Buffon . . . *Mém. Acad. R. Sci.*, 529–548 (published 1749).

Clark-Kennedy, A. E. (1929). *Stephen Hales, D.D., F.R.S. An eighteenth-century biography.* Cambridge.

Clow, A., and N. L. Clow (1952). *The chemical revolution. A contribution to social technology.* London.

Cohen, I. B. (1956). *Franklin and Newton. An inquiry into speculative Newtonian experimental science and Franklin's work in electricity as an example thereof.* Philadelphia, Pa.

——— (1958). Ed. *Isaac Newton's papers and letters on natural philosophy.* Cambridge.

——— (1960). Newton in the light of recent scholarship. *Isis 51*, 489–514.

——— (1963). Pemberton's translation of Newton's *Principia*, with notes on Motte's translation. *Isis 54*, 319–351.

——— (1964). Isaac Newton, Hans Sloane and the Académie Royale des Sciences. In Cohen and Taton (1964), I, 61–116.

——— (1966). Hypotheses in Newton's philosophy. *Physis 8*, 163–184.

Cohen, I. B., and A. Koyré (1961). The case of the missing *tanquam*: Newton, Leibniz and Clarke. *Isis 52*, 555–566.

——— (1962). Newton and the Leibniz-Clarke correspondence. *Archs. int. Hist. Sci. 15*, 52–126.

Cohen, I. B., and R. Taton (1964). Eds. *Mélanges Alexandre Koyré.* 2 vols. Paris.

Colden, C. (1745). *An explication of the first causes of action in matter, and of the cause of gravitation.* New York.

——— (1917–23). *The letters and papers of Cadwallader Colden.* 7 vols. New York.

Coleby, L. M. J. (1938). *The chemical studies of P. J. Macquer.* London.

——— (1952a). John Francis Vigani, first professor of chemistry in the University of Cambridge. *Ann. Sci. 8*, 46–60.

——— (1952b). John Mickleburgh, professor . . . 1718–56. *Ann. Sci. 8*, 165–174.

——— (1952c). John Hadley, fourth professor . . . *Ann. Sci. 8*, 293–301.

Select Bibliography

———(1953). Richard Watson, professor . . . 1764–71. *Ann. Sci. 9*, 101–123.

Coste, P. (1720). *Traité d'optique sur les réflexions, réfractions, inflexions, et couleurs de la lumière. Par M. le Chev. Newton. Traduit de l'Anglois par M. Coste. Sur la seconde édition, augmentée par l'auteur.* 2 vols. Amsterdam.

———(1722). *Traité d'optique . . . seconde édition Françoise, beaucoup plus correcte que la première.* Paris.

Craig, J. (1946). *Newton at the Mint.* Cambridge.

Crook, R. E. (1966). *A bibliography of Joseph Priestley. 1733–1804.* London.

Crosland, M. P. (1959). The use of diagrams as chemical "equations" in the lecture notes of William Cullen and Joseph Black. *Ann. Sci. 15*, 75–90.

———(1963). The development of chemistry in the eighteenth century. *Studies on Voltaire and the Eighteenth Century 24*, 369–441.

Dalton, J. (1793). *Meteorological observations and essays . . .* London.

———(1801). *Elements of English grammar.* London.

———(1802a). Experiments and observations to determine whether the quantity of rain and dew is equal to the quantity of water carried off by the rivers and raised by evaporation, with an inquiry into the origin of springs. *Mem. Proc. Manchester lit. phil. Soc. 5*, 346–372.

———(1802b). New theory of the constitution of mixed aeriform fluids, and particularly of the atmosphere. *Jnl. Nat. Phil. Chem. Arts* [quarto series] *5*, 24.

———(1802c). Experimental essays on the constitution of mixed gases; on the force of steam or vapour from water and other liquids in different temperatures, both in a Torricellian vacuum and in air; on evaporation; and on the expansion of gases by heat. *Mem. Proc. Manchester lit. phil. Soc. 5*, 535–602.

———(1802d). New theory of the constitution of mixed gases elucidated. *Jnl. Nat. Phil. Chem. Arts 3*, 267–271; *Phil. Mag. 14*, 169–173.

———(1805a). Experimental inquiry into the proportion of the several gases or elastic fluids constituting the atmosphere. *Mem. Proc. Manchester lit. phil. Soc.* [2] *1*, 244–258.

———(1805b). On the absorption of gases by water and other liquids. *Mem. Proc. Manchester lit. phil. Soc.* [2] *1*, 271–287.

———(1808–27). *A new system of chemical philosophy.* Part I

Select Bibliography

[1808], part II [1810] and vol. II, part I [1827]. London. (No more published. Facsimile reprint. London, 1953.)

———(1811a). Inquiries concerning the signification of the word particle, as used by modern chemical writers. *Jnl. Nat. Phil. Chem. Arts* 28, 81–88.

———(1811b). Observations on Dr. Bostock's review of the atomic principles of chemistry. *Jnl. Nat. Phil. Chem. Arts* 29, 143–151.

———(1834). *Meteorological observations and essays* . . . 2nd ed. London.

———(1842). *On a new and easy method of analysing sugar.* Manchester.

Daubeny, C. (1831). *An introduction to the atomic theory.* Oxford. (Supplement, London, 1840; 2nd ed., Oxford, 1850.)

Daumas, M. (1953). *Les instruments scientifiques aux XVIIe et XVIIIe siècles.* Paris.

———(1955). *Lavoisier. Théoreticien et expérimentateur.* Paris.

Davis, V. D. (1932). *A history of Manchester College.* London.

Davy, H. (1839–40). *The collected works of Sir Humphry Davy . . . edited by his brother . . .* 9 vols. London.

Debus, A. G. (1965). *The English Paracelsians.* London.

Demachy, J. F. (1774). *Receuil des dissertations physico-chymiques, présentées à différentes Académies . . .* Amsterdam.

Desaguliers, J. T. (1734–44). *A course of experimental philosophy.* 2 vols. London.

Dobbin, L. (1936). A Cullen chemical manuscript of 1753. *Ann. Sci.* 1, 138–156.

Duhem, P. (1902). *Le mixte et la combinaison chimique.* Paris.

Dumas, G. (1936). *Histoire du Journal de Trévoux.* Paris.

Duncan, A. M. (1962). Some theoretical aspects of eighteenth-century tables of affinity. *Ann. Sci.* 18, 177–194 and 217–232.

Dutour, E. F. (1779–82). Expériences relatives à l'adhésion. *Obs. Phys.* 15, 234–252; 16, 85–116; 19, 137–148.

Duveen, D. I., and H. S. Klickstein (1954). *A bibliography to the works of Antoine Laurent Lavoisier . . .* London. (Supplement. London, 1965.)

———(1955). John Dalton's "autobiography." *J. chem. Ed.* 32, 333–334.

Edleston, J. (1850). *Correspondence of Sir Isaac Newton and Professor Cotes, including letters of other eminent men . . .* London.

Elliott, J. (1782). *Elements of the branches of natural philosophy connected with medicine.* London.

Select Bibliography

Farrar, W. V. (1965). Nineteenth-century speculations on the complexity of the chemical elements. *Br. Jnl. Hist. Sci. 8*, 297–323.

Ferguson, A. (1948). Ed. *Natural philosophy through the eighteenth century*. London.

Fontenelle, B. le B. de (1718). Sur les rapports des différentes substances en chimie. *Hist. Acad. R. Sci.*, 35–37 (published 1719).

——(1719). Eloge de M. de Monmort. *Hist. Acad. R. Sci.*, 83–93 (published 1721).

——(1731). Eloge de M. Geoffroy. *Hist. Acad. R. Sci.*, 93–100 (published 1733).

——(1732). Sur l'attraction Newtonienne. *Hist. Acad. R. Sci.*, 112–117 (published 1735).

Forbes, R. I. (1949). Was Newton an alchemist? *Chymia 2*, 27–36.

Fourcroy, A. F. de (1782). *Leçons élémentaires d'histoire naturelle et de chimie* . . . 2 vols. Paris.

——(1784). *Mémoires et observations de chimie* . . . Paris.

——(1787). *Principes de chimie.* 2 vols. Paris.

——(1788). *Elements of natural history and of chemistry* . . . 4 vols. London (translation of 2nd French ed. by W. Nicholson).

——(1800). *Système des connaissances chimiques* . . . 10 vols. Paris.

——(1804). *A general system of chemical knowledge* . . . Translated by W. Nicholson. 11 vols. London.

——(1808). *Encyclopédie méthodique. Chimie, pharmacie et métallurgie* . . . Vol. 5. Paris.

Fox, R. (1968). The background to the discovery of Dulong and Petit's Law. *Br. Jnl. Hist. Sci. 4*, 1–22.

Freind, J. (1703). *Emmenologia: in qua fluxis muliebris menstrui phaenomena, periodi, vitia cum modendi methodo, ad rationes mechanicus exiguntur.* Oxford (English translation, London, 1729).

——(1709). *Praelectiones chymicae, in quibus omnes fere operationes chymicae ad vera principia & ipsius naturae leges rediguntur, Ann. 1704. Oxonii, in museo Ashmoleano habitae.* London.

——(1710). *Praelectiones chymicae, in quibus omnes fere operationes chymicae ad vera principia et ipsius natura leges rediguntur, Oxonii habitae.* Amsterdam.

——(1712a). *Chymical lectures . . . Englished by J. M. To which is added, an appendix, containing the account given of this book in the Lipsick Acts, together with the author's remarks thereon.* London.

——(1712b). Praelectionum chymicarum vindiciae, in quibus objectiones, in Actis Lipsiensibus anno 1710, mense Septembri, contra vim materiae attractricam allatae, duluuntur. *Phil. Trans. R. Soc.,* no. 331, 330–342 (Latin version of Freind's additional remarks in 1712a).

Freund, I. (1904). *The study of chemical composition. An account of its method and historical development. With illustrative quotations.* Cambridge.

Fric, R. (1955–   ). Ed. *Oeuvres de Lavoisier,* tome VII, *Correspondance,* Fascicules I— , 1763–

Gay, P. (1967). *The Enlightenment: an interpretation. The rise of modern paganism.* London.

Geoffroy, E. F. (1718). Table des différents rapports observés en chimie entre différentes substances. *Mém. Acad. R. Sci.,* 202–212 (published 1719).

Geoghegan, D. (1957). Some indications of Newton's attitude toward alchemy. *Ambix 6,* 102–106.

Gibbs, F. W. (1951a). Robert Dossie (1717–1777) and the Society of Arts. *Ann. Sci. 7,* 149–172.

———(1951b). Peter Shaw and the revival of chemistry. *Ann. Sci. 7,* 211–237.

———(1952). William Lewis, M. B., F. R. S. (1708–1781). *Ann. Sci. 8,* 122–151.

———(1957). Boerhaave's chemical writings. *Ambix 6,* 117–135.

———(1961). Itinerant lecturers in natural philosophy. *Ambix 8,* 111–117.

———(1965a). Bryan Higgins & his circle. *Chem. in Br. 2,* 60–65.

———(1965b). *Joseph Priestley. Adventurer in science and champion of truth.* London.

Gibbs, F. W., and W. A. Smeaton (1961). Thomas Beddoes at Oxford. *Ambix 9,* 47–49.

Gillispie, C. C. (1959). The *Encyclopédie* and the Jacobin philosophy of science; a study in ideas and consequences. In Clagett (1959), 255–290.

Gough, J. B. (1968). Lavoisier's early career in science. An examination of some new evidence. *Br. Jnl. Hist. Sci. 4,* 52–57.

'sGravesande, W. J. (1720–21). *Physices elementa mathematica, experimentis confirmata: sive introductio ad philosophiam Newtonianam.* 2 vols. Leyden.

———(1720a). *Mathematical elements of natural philosophy, confirmed by experiments, or an introduction to Sir Isaac Newton's philosophy . . . translated into English by J. T.* 2 vols. London.

———(1720b). *Mathematical elements of physicks, prov'd by ex-*

Select Bibliography

periments: being an introduction to Sir Isaac Newton's philosophy . . . *Made English* . . . *Revis'd and corrected by Dr. John Keill* . . . London.

——(1735).* *An explanation of the Newtonian philosophy, in lectures read to the youth of the University of Leyden* . . . *translated* . . . *by a Fellow of the Royal Society*. London.

Gray, G. J. (1907). *A bibliography of the works of Sir Isaac Newton*. 2nd ed. revised and enlarged. Cambridge.

Green, R. (1711). *A demonstration of the truth and divinity of the Christian religion* . . . Cambridge.

——(1712a). *The principles of natural philosophy* . . . Cambridge.

——(1712b). *A taste of philosophical fanaticism; in some speculations upon the first four chapters of Mr. Green's "Principles of natural philosophy." By a gentleman of the University of Gratz in Germany*. London.

——(1727). *The principles of the philosophy of the expansive and contractive forces, or an inquiry into the principles of the modern philosophy* . . . Cambridge.

Greenaway, F. (1958). The biographical approach to John Dalton. *Mem. Proc. Manchester lit. phil. Soc.*, 100.

——(1966). *John Dalton and the atom*. London.

Gregory, D. (1702). *Astronomiae physicae & geometricae elementa*. Oxford.

Gregory, J. C. (1931). *A short history of atomism from Democritus to Bohr*. London.

Guerlac, H. (1951). The Continental reputation of Stephen Hales. *Archs. int. Hist. Sci. 4*, 393–404.

——(1959). Some French antecedents of the chemical revolution. *Chymia 5*, 73–112.

——(1961a). *Lavoisier—the crucial year. The background and origin of his first experiments on combustion in 1772*. Ithaca, New York.

——(1961b). Some Daltonian doubts. *Isis 52*, 544–554.

——(1961c). Quantification in chemistry. *Isis 52*, 194–214.

——(1963a). *Newton et Epicure*. Conférence donnée au Palais de la Découverte, le 2 Mars 1963. Paris.

——(1963b). Francis Hauksbee, expérimentateur au profit de Newton. *Archs. int. Hist. Sci. 16*, 113–128.

——(1964). Sir Isaac and the ingenious Mr. Hauksbee. In Cohen and Taton (1964), I, 228–253.

——(1965). Where the statue stood: divergent loyalties to Newton in the eighteenth century. In Wasserman (1965), 317–334.

——— (1967). Newton's optical ether. *Notes Rec. R. Soc. Lond.* 22, 45–57.
Guyton (1772). *Digressions académiques, ou essais sur quelques sujets* ... Dijon.
——— (1773). Expériences sur l'attraction ou la répulsion de l'eau & des corps huileux, pour vérifier l'exactitude de la méthode par laquelle le Docteur Taylor estime la force d'adhésion des surfaces ... *Obs. Phys.* 1, 168–173.
——— (1776). Article "Affinité" in the *Supplément à l'Encyclopédie*. 4 vols. Amsterdam.
——— (1777–78). *Elémens de chymie, théorique et pratique, rédigés dans un nouvel ordre* ... 3 vols. Dijon.
——— (1789). *Encyclopédie méthodique. Chymie, pharmacie et métallurgie* ... Vol. 1. Paris.
Hahn, R. (1967). *Laplace as a Newtonian scientist*. (William Andrews Clark Memorial Library Publications). Los Angeles, California.
Hales, S. (1727). *Vegetable staticks: or, an account of some statical experiments on the sap in vegetables* ... *Also, a specimen of an attempt to analyse the air, by a great variety of chymiostatical experiments* ... London.
——— (1733). *Haemastaticks; or an account of some hydraulick and hydrostatical experiments made on the blood and blood-vessels of animals. Also an account of some experiments on stones in the kidneys and bladder* ... London.
Hall, A. R., and M. B. Hall (1958). Newton's chemical experiments. *Archs. int. Hist. Sci.* 11, 113–152.
——— (1959). Newton's mechanical principles. *Jnl. Hist. Ideas* 20, 167–178.
——— (1960). Newton's theory of matter. *Isis* 51, 131–144.
——— (1962). *Unpublished scientific papers of Isaac Newton*. Cambridge.
[Hall], M. Boas (1952). The establishment of the mechanical philosophy. *Osiris* 10, 413–541.
——— (1958). *Robert Boyle and seventeenth-century chemistry*. Cambridge.
——— (1959). Structure of matter and chemical theory in the seventeenth and eighteenth centuries. In Clagett (1959), 499–514.
——— (1965). *Robert Boyle on natural philosophy. An essay with selections from his writings*. Bloomington, Indiana.
Halley, E. (1693). An account of the measure of the thickness of gold upon gilt-wire, together with a demonstration of the exceeding

Select Bibliography

minuteness of the atoms or constituent particles of gold . . . *Phil. Trans. R. Soc.,* no. 194, 540–542.

Hanks, L. (1966). *Buffon avant "l'Histoire Naturelle."* Paris.

Hannequin, A. (1895). *Essai critique sur l'hypothèse des atomes dans la science contemporaine.* Paris.

Hans, N. A. (1951). *New trends in education in the eighteenth century.* London.

Harden, A. (1915). John Dalton's lectures and lecture illustrations. *Mem. Proc. Manchester lit. phil. Soc. 59,* no. 12.

See also Roscoe and Harden.

Harpe, J. de la (1937–41). Le *Journal des Savants* et l'Angleterre, 1702–1789. *University of California publications in modern philology 20,* 289–520.

Harris, J. (1704). *Lexicon technicum: or, an universal English dictionary of arts and sciences* . . . Vol. 1. London.

———(1710). *Lexicon technicum: or, an universal English dictionary of arts and sciences* . . . Vol. 2. London.

———(1744). *A supplement to Dr. Harris's dictionary . . . by a Society of Gentlemen.* London.

Harrison, J. A. (1957). "Blind Henry Moyes." An excellent lecturer in philosophy. *Ann. Sci. 13,* 109–125.

Hauksbee, F. (1708a). An account of some experiments, touching the electricity and light producible on the attrition of several bodies. *Phil. Trans. R. Soc.,* no. 315, 87–92.

———(1708b). Several experiments touching the seeming spontaneous ascent of water. *Phil. Trans. R. Soc.,* no. 319, 258–268.

———(1709). *Physico-mechanical experiments on various subjects. Containing an account of several surprising phenomena touching light and electricity, producible on the attrition of bodies.* London.

Hazard, P. (1953). *The European mind (1680–1715).* Translated by J. L. May. London.

Heathcote, N. H. de V., and D. McKie (1935). *The discovery of specific and latent heats.* London.

———(1958). William Cleghorn's *De Igne* (1779). *Ann. Sci. 14,* 1–82.

Helsham, R. (1739). *A course of lectures in natural philosophy . . . Published by Bryan Robinson.* London.

Henry, W. (1802). A review of some experiments which have been supposed to disprove the materiality of heat. *Mem. Proc. Manchester lit. phil. Soc. 5,* 603–621.

———(1803). Experiments on the quantity of gases absorbed by

Select Bibliography

water, at different temperatures, and under different pressures. *Phil. Trans. R. Soc. 93*, 29–42 and 274–276.

——(1804). Illustrations of Mr. Dalton's theory of the constitution of mixed gases . . . *Jnl. Nat. Phil. Chem. Arts 8*, 297–301.

Henry, W. C. (1854). *Memoirs of the life and scientific researches of John Dalton.* London.

Hesse, M. B. (1961). *Forces and fields. The concept of action at a distance in the history of physics.* London.

Higgins, B. (1776). *A philosophical essay concerning light.* Vol. 1. London (no more published).

——(1786). *Experiments and observations relating to acetous acid* . . . London.

——(1795). *Minutes of a society for philosophical experiments and conversations.* London.

Higgins, W. (1789). *A comparative view of the phlogistic and antiphlogistic theories* . . . London.

——(1791). *A comparative view of the phlogistic and antiphlogistic theories* . . . 2nd ed. London (reproduced in facsimile in Partington and Wheeler [1960]).

——(1814). *Experiments and observations on the atomic theory and electrical phenomena.* London (reproduced in facsimile in Partington and Wheeler [1960]).

Hiotzeberg, M. (1772). Dissertation sur la cause de l'attraction des corps . . . *Obs. Phys., Introduction 1*, 527–534.

Hiscock, W. G. (1937). *David Gregory, Isaac Newton and their circle.* Oxford.

Hoefer, F. (1842–43). *Histoire de la chimie.* 2 vols. Paris.

Holmes, F. L. (1962). From elective affinities to chemical equilibria; Berthollet's law of mass action. *Chymia 8*, 105–145.

Homberg, W. (1702). Essais de chimie. Des principes de la chimie en général. *Mém. Acad. R. Sci.*, 33–52 (published 1704).

——(1705). Suite des essais de chimie. Article troisième. Du souphre principe. *Mém. Acad. R. Sci.*, 88–96 (published 1706).

Hooke, R. (1661). *An attempt for the explication of the phaenomena observable in an experiment published by the Honourable Robert Boyle, Esq; in the XXXV experiment of his epistolical discourse touching the aire, in confirmation of a former conjecture made by R. Hooke.* London.

Horne, G. (1753). *A fair, candid and impartial state of the case between Sir Isaac Newton and Mr. Hutchinson, in which is shewn, how far a system of physics is capable of mathematical demonstration* . . . Oxford (2nd ed. London, 1799 *).

Select Bibliography

———(1799). *The works of George Horne . . . to which are prefixed memoirs of his life . . . by W. Jones.* 6 vols. London.

Hoskin, M. A. (1951). 'Mining all within': Clarke's notes to Rohault's *Traité de Physique. The Thomist* 24, 353–363.

Hughes, A. (1951–55). Science in English encyclopedias (1704–1875). *Ann. Sci.* 7, 340–370; *8,* 323–367; *9,* 233–264; *11,* 74–92.

Hutchinson, J. (1724).* *Moses Principia. I. Of the invisible parts of matter. Of the motion of the visible forms and of their dissolutions and reformations.* London. (Part II. *Of the circulation of the heavens.* London, 1728.)

———(1748). *The philosophical and theological works of John Hutchinson . . .* 12 vols. London.

———(1753).* *An abstract from the works of J. Hutchinson . . .* Edinburgh (and 1755, London).

Hutton, C. (1795). *A mathematical and philosophical dictionary.* 2 vols. London.

Ihde, A. J. (1964). *The development of modern chemistry.* London.

Jacquot, J. (1953). Sir Hans Sloane and French men of science. *Notes Rec. R. Soc. Lond.* 10, 85–98.

———(1954). *Le naturaliste Sir Hans Sloane (1660–1753) et les exchanges scientifiques entre la France et l'Angleterre.* (Libraire du Palais de la Découverte) Paris.

Jammer, M. (1957). *Concepts of force: a study in the foundations of dynamics.* Cambridge, Mass.

———(1961). *Concepts of mass in classical and modern physics.* Cambridge, Mass.

Johnston, G. A. (1930). *Berkeley's commonplace book. Edited with an Introduction, Notes and Index.* London.

Jones, R. M. (1921). *The later periods of Quakerism.* 2 vols. London.

Jones, W. (1762). *An essay on the first principles of natural philosophy.* Oxford.

———(1781). *Physiological disquisitions, or discourses on the natural philosophy of the elements.* London.

———(1801). *The theological, philosophical and miscellaneous works of . . . William Jones . . . to which is prefixed a short account of his life and writings.* 12 vols. London.

Jones, W. P. (1966). *The rhetoric of science. A study of scientific ideas and imagery in eighteenth-century English poetry.* London.

Kargon, R. (1965). W. R. Hamilton and Boscovichean atomism. *Jnl Hist. Ideas* 26, 137–140.

Select Bibliography

———(1966). *Atomism in England from Hariot to Newton.* Oxford.

Keill, James (1698). *The anatomy of the human body, abridged.* London.

———(1708a). *An account of animal secretion, the quantity of blood in the humane body, and muscular motion.* London.

See also Lemery (1698).

Keill, John (1702). *Introductio ad veram physicam, seu lectiones physicas, habitas in schola naturalis philosophiae academiae Oxoniensis* . . . Oxford (2nd ed. Oxford, 1705. English translation of the 1719 fourth [London] edition as Keill [1720]).

———(1708b). Epistola ad cl. virum Gulielmum Cockburn, medicinae doctorem. In qua leges attractionis alinquae physices principia traduntur. *Phil. Trans. R. Soc.,* no. 315, 97–110.

———(1708c). Epistola ad clarissimum virum Edmundum Halleium Geometriae Professorem Savilianum, de legibus virium centripatarum. *Phil. Trans. R. Soc.,* no. 317, 174–188.

———(1714). Theoremata quaedam infinitam materiae divisibilitatem spectantia, quae ejusdem raritatem & tenuom compositionem demonstrant, quorum ope plurimae in physica tolluntur difficultates. *Phil. Trans. R. Soc.,* no. 339, 82–86.

———(1715). *Introductio ad veram physicam* . . . 3rd ed. Oxford.

———(1720). *An introduction to natural philosophy, or philosophical lectures read in the University of Oxford. A.D. 1700* . . . London.

———(1725). *Introductiones ad veram physicam et veram astronomiam. Quibus accedunt trigonometria. De viribus centralis. De Legibus attractionis.* [With introductory note by W. J. 'sGravesande] Leyden.

Kent, A. (1950). Ed. *An eighteenth-century lectureship in chemistry. Essays and bicentenary addresses relating to the chemistry department of Glasgow University.* Glasgow.

Kerker, M. (1955). Herman Boerhaave and the development of pneumatic chemistry. *Isis 46,* 36–49.

Kiernan, C. (1968). Science and the Enlightenment in eighteenth-century France. *Studies on Voltaire and the Eighteenth Century 59.*

King, P. (1968). Bishop Wren and the suppression of the Norwich lecturers. *Camb. Hist. Jnl. 11,* 237–254.

Kirwan, R. (1781). Experiments and observations on the specific gravities and attractive powers of various saline substances. *Phil. Trans. R. Soc. 71,* 7–41.

Select Bibliography

——(1782). Continuation of the experiments . . . *Phil. Trans. R. Soc.* 72, 179–236.

——(1783). Conclusion of the experiments . . . *Phil. Trans. R. Soc.* 73, 15–84.

——(1791). Of the strengths of acids, and of the proportion of ingredients in neutral salts. *Trans. R. Ir. Acad.* 4, 3–84.

——(1800). Additional observations on the proportion of real acid in three antient known mineral acids, and on the ingredients in various neutral salts . . . *Trans. R. Ir. Acad.* 7, 163–304.

Knight, D. M. (1967). *Atoms and elements. A study of theories of matter in England in the nineteenth century.* London.

Knight, G. (1754). *An attempt to demonstrate, that all the phaenomena in nature may be explained by two simple active principles, attraction and repulsion* . . . London (original ed. 1748).

——(1758). *A collection of some papers formerly published in the "Philosophical Transactions," relative to Dr. Knight's magnetical bars.* London.

Kopp, H. (1843–47). *Geschichte der Chemie.* 4 vols., Braunschweig. (Reprinted Hildesheim, 1966.)

Koyré, A. (1955). Pour une édition critique des œuvres de Newton. *Revue Hist. Sci. Applic.* 3, 19–37.

——(1957). *From the closed world to the infinite universe.* Baltimore, Md.

——(1960a). Etudes Newtoniennes. I—Les regulae philosophandi. *Archs. int. Hist. Sci.* 13, 3–14.

——(1960b). Etudes Newtoniennes. II—Les queries de l'Optique. *Archs. int. Hist. Sci.* 13, 15–29.

——(1961). Etudes Newtoniennes. III—Attraction, Newton, and Cotes. *Archs. int. Hist. Sci.* 14, 225–236.

See also Cohen and Koyré (1961 and 1962).

Kranzberg, M., and C. W. Pursell, Jr. (1967). *Technology in western civilization.* 2 vols. London.

Kronick, D. A. (1962). *A history of scientific and technical periodicals . . . 1665–1790.* New York.

Kuhn, A. J. (1961). Glory or gravity. Hutchinson vs. Newton. *Jnl. Hist. Ideas* 22, 303–322.

Kuhn, T. S. (1952). Robert Boyle and structural chemistry in the seventeenth century. *Isis* 43, 12–36.

——(1962). *The structure of scientific revolutions.* London.

Lange, F. A. (1877–81). *The history of materialism, and criticism of its present importance.* Translated by E. C. Thomas. 3 vols. London.

Select Bibliography

Laplace, P. S. (1878–1912). *Oeuvres complètes de Laplace* . . . 14 vols. Paris.

Lasswitz, K. (1890). *Geschichte der Atomistik von Mittelalter bis Newton*. 2 vols. Hamburg.

Lavoisier, A. L. (1770). Premier mémoire sur la nature de l'eau, et sur les expériences par lesquelles on a prétendu prouver la possibilité de son changement en terre . . . *Mém. Acad. R. Sci.,* 73–82 (published 1773).

——(1782). Mémoire sur l'affinité du principe oxygine avec les différentes substances auxquelles il est susceptible de s'unir. *Mém. Acad. R. Sci.,* 530–540 (published 1785).

——(1787). *Méthode de nomenclature chimique, proposée par MM. de Morveau, Lavoisier, Berthollet, & de Fourcroy*. Paris.

——(1790). *Elements of chemistry in a new systematic order, containing all the modern discoveries.* Translated from the French by Robert Kerr. Edinburgh. (Facsimile reprint, New York, 1965.)

Lecky, W. E. H. (1878–90). *History of England in the eighteenth century*. 8 vols. New York.

Lemery, N. (1675). *Cours de chymie*. Paris (11th ed., Paris, 1730).

——(1698). *A course of chymistry . . . translated by James Keill from the eighth French edition* . . . London.

Lilley, S. (1950). Nicholson's Journal (1797–1813). *Ann. Sci.* 6, 78–101.

Limbourg, J. P. de (1761). *Dissertation . . . sur les affinités chymiques*. Liège.

Lindeboom, G. A. (1959). *Bibliographia Boerhaaviana (Analecta Boerhaaviana,* Vol. 1). Leyden.

——(1962–64). *Boerhaave's correspondence, Parts 1 and 2*. Leyden.

——(1963). Pitcairne's Leyden interlude described from the documents. *Ann. Sci. 19,* 273–284.

Loewenfeld, K. (1913). Contributions to the history of science (Period of Priestley-Lavoisier-Dalton) based on autograph documents. *Mem. Proc. Manchester lit. phil. Soc.* 57, no. 19.

Lonsdale, H. (1874). *The worthies of Cumberland. John Dalton.* London.

McCosh, J. (1875). *The Scottish Philosophy, biographical, expository, critical, from Hutchinson to Hamilton.* London.

McGuire, J. E. (1966). Body and void and Newton's *De Mundi Systemate:* some new sources. *Archs. Hist. exact Sci.* 3, 206–248.

Select Bibliography

———(1967). Transmutation and immutability: Newton's doctrine of physical qualities. *Ambix 14*, 69–95.

McGuire, J. E., and P. M. Rattansi (1966). Newton and the "pipes of Pan." *Notes Rec. R. Soc. Lond. 21*, 108–143.

Mach, E. (1907). *The science of mechanics. A critical and historical account of its development . . . Translated . . . by J. McCormack.* 3rd ed. Chicago.

McKie, D. (1936). On Thos. Cochrane's Ms. notes of Black's chemical lectures, 1767–68. *Ann. Sci. 1*, 101–110.

———(1942). Some notes on Newton's chemical philosophy . . . *Phil. Mag.* [7] *33*, 847–870.

———(1951). Mr. Warltire, a good chymist. *Endeavour 10*, 46–49.

———(1957). Bernard le Bovier de Fontenelle, F.R.S. (1657–1757). *Notes Rec. R. Soc. Lond. 12*, 193–200.

———(1959). On some Ms. copies of Black's chemical lectures—II. *Ann. Sci. 15*, 65–73.

———(1960). On some letters of Joseph Black and others. *Ann. Sci. 16*, 129–170.

———(1962a). On some Ms. copies of Black's chemical lectures—III. *Ann. Sci. 18*, 1–9.

———(1962b). Black's chemical lectures—IV. *Ann. Sci. 18*, 87–97.

———(1965). Black's chemical lectures—V. *Ann. Sci. 21*, 209–255.

———(1966). Ed. *Thomas Cochrane. Notes from Doctor Black's lectures on chemistry 1767/8.* Wilmslow, Cheshire. (Privately published by Imperial Chemical Industries.)

See also Heathcote and McKie (1935 and 1958).

McKie, D., and J. R. Partington (1937–39). Historical studies on the phlogiston theory. *Ann. Sci. 2*, 361–404; *3*, 1–58 and 337–371; *4*, 113–149.

McLachlan, H. (1931). *English education under the Test Acts; being the history of the Nonconformist Academies, 1662–1820.* Manchester.

———(1945). John Dalton and Manchester. *Mem. Proc. Manchester lit. phil. Soc. 86*, 165–177.

McMullen, E. (1963). Ed. *The concept of matter.* Notre Dame, Indiana.

Macquer, P. J. (1749). *Elémens de chymie théorique.* Paris.

———(1751). *Elémens de chymie pratique.* 2 vols. Paris.

———(1758). *Elements of the theory and practise of chemistry. Translated . . . by A. Reid.* 2 vols. London.

———(1766). *Dictionnaire de chymie . . .* [Published anonymously.] 2 vols. Paris.

Select Bibliography

———(1771). *A dictionary of chemistry . . . translated from the French.* 2 vols. London.

Mandelbaum, M. (1964). *Philosophy, science and sense perception. Historical and critical studies.* Baltimore, Md.

Manuel, F. E. (1963). *Isaac Newton, Historian.* Cambridge, Mass.

———(1968). *A portrait of Isaac Newton.* Cambridge, Mass.

Marchand, P. (1758–59). *Dictionnaire historique.* 2 vols. La Haye.

Martin, B. (1735). *The philosophical grammar; being a view of the present state of experimented physiology, or natural philosophy.* London. (7th ed. London, 1769. French translation as *Grammaires des sciences philosophiques . . .* Paris, 1749).

———(1747). *Philosophia Britannica: or a new and comprehensive system of the Newtonian philosophy . . .* 2 vols. London.

———(1749). *A panegyrick on the Newtonian philosophy . . .* London.

———(1751). *A plain and familiar introduction to the Newtonian philosophy . . . designed for the use of such gentlemen and ladies as would acquire a competent knowledge of this science, without mathematical learning . . .* London.

Maupertuis, P. L. M. de (1732). Sur les loix de l'attraction. *Mém. Acad. R. Sci.,* 343–362 (published 1735).

Mead, R. (1702). *A mechanical account of poisons in several essays.* London.

———(1704). *De imperio solis ac lunae in corpora humana, et morbis inde oriundis.* London.

———(1712). *Of the power and influence of the sun and moon on humane bodies; and of the diseases that rise from thence.* London.

———(1755). [Anon.] *Authentic memoirs of the life of Richard Mead, M.D.* London.

Meldrum, A. N. (1911). The development of the atomic theory. *Mem. Proc. Manchester lit. phil. Soc. 54,* no. 7; *55,* nos. 3, 4, 5, 6, 19, and 22.

Metzger, H. (1923). *Les doctrines chimiques en France du début du XVIIe à la fin du XVIIIe siècle. Première partie.* Paris.

———(1930). *Newton, Stahl, Boerhaave et la doctrine chimique.* Paris.

———(1935). *La philosophie de la matière chez Lavoisier.* Paris.

———(1938). *Attraction universelle et religion naturelle chez quelques commentateurs anglais de Newton.* Paris.

Meyer, E. von (1891). *A history of chemistry from earliest times to the present day . . .* Translated by George McGowan, London.

Select Bibliography

Meyerson, E. (1930). *Identity and reality. Authorised translation by K. Loewenberg.* London.

Middleton, W. E. K. (1965). *A history of the theories of rain* . . . London.

Mitford, N. (1957). *Voltaire in love.* London.

More, L. T. (1934). *Isaac Newton. A biography.* London.

Morgan, B. T. (1929). *Histoire du* "Journal des Savants." Paris.

Mornet, D. (1911). *Les sciences de la nature en France, au XVIIIe siècle.* Paris.

Muir, M. M. P. (1906). *A history of chemical theories and laws.* London.

Mullet, C. F. (1940). *The letters of George Cheyne to the Countess of Huntingdon.* California.

Munby, A. N. L. (1949). A catalogue of the manuscripts and printed books in the Sir Isaac Newton collection . . . bequeathed . . . to King's College, Cambridge. (Typescript. Copies in King's College and the University Library, Cambridge.)

———(1953). The Keynes collection of the works of Sir Isaac Newton. *Notes Rec. R. Soc. Lond.* **10**, 40–50.

Musschenbroek, P. van (1729). *Physicae experimentales, et geometricae, de magnete, tuborum capillarium vitreorumque, speculorum attractione, magnitudine terrae, cohaerentia corporum firmorum dissertationes* . . . Leyden.

———(1734).* *Elementa physicae, conscripta in usos academicos.* Leyden.

———(1739). *Essai de physique* . . . *traduit du Hollandois par Mr. P. Massuet.* 2 vols. Leyden.

———(1744). *The elements of natural philosophy. Chiefly intended for the use of students in universities* . . . *translated* . . . *by J. Colson.* 2 vols. London (translation of [1734]).

———(1762).* *Introductio ad philosophiam naturalem.* 2 vols. Leyden.

———(1769). *Cours de physique experimentale et mathematique* . . . 3 vols. Paris (translation of [1762], by Sigaud de la Fond).

Musson, A. E., and E. Robinson (1960). Science and industry in the late eighteenth century. *Econ. Hist. Rev.* [2] **13**, 222–244.

Nash, L. K. (1956). The origin of Dalton's chemical atomic theory. *Isis* **47**, 101–116.

Neville, R. G. (1960). Unrecorded Daltoniana . . . *Ambix* **8**, 42–45.

Newton, I. (1687). *Philosophiae naturalis principia mathematica.* London. (Facsimile reprint, London, 1953.)

———(1704). *Opticks: or, a treatise of the reflexions, refractions,*

Select Bibliography

inflexions and colours of light. London. (Facsimile reprint, Brussels, 1966.)

——(1706). *Optice: sive de reflexionibus, refractionibus, inflexionibus & coloribus lucis libri tres. Latine reddidit S. Clarke* . . . London.

——(1713). *Philosophiae naturalis principia mathematica. Editio secunda auctior et emendatior.* Cambridge.

——(1715). An account of the Commercium Epistolicum. *Phil. Trans. R. Soc. Lond.,* no. 342, 173–224.

——(1718). *Opticks . . . The second edition, with additions.* 2nd printing. (1st printing, 1717) London.

——(1721). *Opticks . . . The third edition, corrected.* London.

——(1726). *Philosophiae naturalis principia mathematica. Editio tertia aucta & emendata.* London.

——(1729). *The mathematical principles of natural philosophy. Translated into English by Andrew Motte.* 2 vols. London.

——(1730). *Opticks . . . The fourth edition, corrected.* London.

——(1756). *Four letters from Sir Isaac Newton to Doctor Bentley* . . . London. (Facsimile reproduction in Cohen [1958].)

Nichols, J. (1817–58). *Illustrations of the literary history of the eighteenth century.* 8 vols. London.

Nicholson, C. (1861). *Annals of Kendal.* 2nd ed. London.

Nicholson, M. (1946). *Newton demands the muse. Newton's "Opticks" and the eighteenth-century poets.* Princeton, N. J.

Nicholson, W. (1788). *An introduction to natural philosophy.* 3rd ed. Philadelphia, Pa.

——(1790). *First principles of chemistry.* London.

——(1795). *A dictionary of chemistry.* 2 vols. London.

Paris, J. A. (1831). *The life of Sir Humphry Davy.* 2 vols. London.

Partington, J. R. (1937). *A short history of chemistry.* London.

——(1939). The origins of the atomic theory. *Ann. Sci. 4,* 245–282.

——(1961–64). *A history of chemistry.* Vols. 2–4. (Vol. 1 not published.) London.

See also McKie and Partington (1939).

Partington, J. R., and T. S. Wheeler (1960). *The life and work of William Higgins* . . . London.

Pearson, G. (1799). *A translation of the tables of chemical nomenclature . . . To which are subjoined . . . tables of the precise forces of chemical attractions . . .* 2nd ed. London.

——(1806). *Heads and notes of a course of chemical lectures.* London.

Select Bibliography

Pelseneer, J. (1952). Une lettre inédite de Buffon à Guyton de Morveau à propos du phlogistique. In *Buffon*, by L. Bertin *et al.* (Paris, 1952); 133–136.

Pemberton, H. (1728). *A view of Sir Isaac Newton's philosophy.* London.

——(1771). *A course of chemistry . . . formerly given by the late learned Doctor Henry Pemberton . . . now first published from the author's manuscript by James Wilson, M.D.* London.

Pitcairne, A. (1695). *Apollo mathematicus: or the art of curing diseases by the mathematicks, according to the principles of Dr. Pitcairne . . .* [Edinburgh].

——(1715). *The works of Dr. Archibald Pitcairn . . . with some account of the author's life, prefixed . . .* London.

Plummer, A. (1754a). Remarks on chemical solutions and precipitations . . . In *Essays and observations, physical and literary. Read before a Society in Edinburgh, and published by them.* 3 vols. (Edinburgh, 1754–71); I, 284–313.

——(1754b). Experiments on neutral salts . . . In *Essays and observations, physical and literary. Read before a Society in Edinburgh, and published by them.* 3 vols. (Edingburgh, 1754–71); I, 315–338.

Portsmouth (1888). *A catalogue of the Portsmouth collection of books and papers written by or belonging to Sir Isaac Newton . . .* Cambridge.

Priestley, J. (1772). *The history and present state of discoveries relating to vision, light and colours.* 2 vols. London.

——(1774). *An examination of Dr. Reid's inquiry into the human mind on the principles of common sense . . .* London.

——(1777). *Disquisitions relating to matter and spirit.* London.

——(1778). *A free discussion of the doctrines of materialism and philosophical necessity . . .* London.

——(1779–86). *Experiments and observations relating to various branches of natural philosophy . . .* 3 vols. London.

——(1793). *Experiments on the generation of air from water . . .* London. (Reprinted from *Phil. Trans. R. Soc. Lond.*)

——(1794). *Heads of lectures on a course of experimental philosophy, particularly including chemistry . . .* Dublin.

Ramsay, W. (1918). *The life and letters of Joseph Black . . .* London.

Rappaport, R. (1960). G. F. Rouelle: an eighteenth-century chemist and teacher. *Chymia 6,* 68–101.

——(1961). Rouelle and Stahl—the phlogistic revolution in France. *Chymia 7,* 73–102.

Select Bibliography

Rattansi, P. M. (1963). Paracelsus and the Puritan revolution. *Ambix 11*, 24–32. See also McGuire and Rattansi (1966).

Reid, T. (1785). *Essays on the intellectual powers of man*. Edinburgh.

Richter, J. B. (1791–1802). *Ueber die neuem Gegenstände der Chymie*. 11 vols. Bresslau, Hirschberg and Lisza.

——— (1792–94). *Anfangsgründe der Stochyometrie* . . . 3 vols. Bresslau and Hirschberg.

Rigaud, S. P. (1838). *Historical essay on the first publication of Sir Isaac Newton's "Principia."* Oxford.

Rigaud, S. P., and S. J. Rigaud (1841). *Correspondence of scientific men of the seventeenth century* . . . 2 vols. Oxford.

Robinet, A. (1957). *Correspondance Leibniz-Clarke. Présentée d'après les manuscripts originaux* . . . Paris.

Robinson, B. (1704). *New elements of conick sections . . . translated from the French treatise of Mr. de la Hire by B. Robinson*. London.

——— (1732). *A treatise of the animal oeconomy*. Dublin.

——— (1737). *A short essay on coin*. Dublin.

——— (1743). *A dissertation of the aether of Sir Isaac Newton*. Dublin (and London, 1747).

——— (1745). *Sir Isaac Newton's account of the aether, with some additions by way of appendix*. Dublin.

——— (1746). *An appendix to the dissertation on the aether*. (Dated Dec. 1, 1746, and paginated continuously with Robinson [1743]. No place or year of issue given.)

——— (1747). *A dissertation on the food and discharges of human bodies*. Dublin (and London, 1748).

——— (1752). *Observations on the virtues and operations of medicines*. London.

Robison, J. (1804). *Elements of mechanical philosophy*. Vol. 1. Edinburgh. (No more published.)

——— (1822). *A system of mechanical philosophy . . . edited by D. Brewster*. 4 vols. Edinburgh.

Roscoe, H. E. (1895). *John Dalton and the rise of modern chemistry*. London.

Roscoe, H. E., and A. Harden (1896). *A new view of the origin of Dalton's atomic theory*. London.

Rowbottom, M. (1968). The teaching of experimental philosophy in England, 1700–1730. *Act XIe Cong. int. Hist. Sci. 4*, 46–63. Warsaw.

Select Bibliography

Rowning, J. (1735). *A compendious system of natural philosophy.* 2 vols. Cambridge.

Rutherforth, T. (1748). *A system of natural philosophy* . . . 2 vols. Cambridge.

Sabra, A. I. (1967). *Theories of light. From Descartes to Newton.* London.

Scheele, C. W. (1786). *The chemical essays of Charles William Scheele translated from the transactions of the Academy of Sciences at Stockholm.* London (reissued in 1901: all quotations from the reissued edition).

Schofield, R. E. (1953). John Wesley and science in eighteenth-century England. *Isis 44,* 331–340.

———(1963). *The Lunar Society of Birmingham. A social history of provincial science and industry in eighteenth-century England.* Oxford.

———(1966). Joseph Priestley, natural philosopher. *Ambix 14,* 1–15.

———(1967). *A scientific autobiography of Joseph Priestley (1733–1804). Selected scientific correspondence edited with commentary.* Cambridge, Mass.

Scott, J. F. (1967–). *The correspondence of Isaac Newton.* Vols. 4– , 1694– . Cambridge.

Senac, J. B. (1723). *Nouveau cours de chymie, suivant les principes de Newton & de Stahl.* 2 vols. Paris (2nd ed., 2 vols., Paris, 1737).

Shaw, P. (1723). *A treatise of incurable diseases.* London.

———(1725). *The philosophical works of the Honourable Robert Boyle . . . abridged, methodized, and . . . illustrated with notes . . . by Peter Shaw, M.D.* 3 vols. London.

———(1731). *Scheme for a course of philosophical chemistry: or the application of chemical experiments to the improvement of natural knowledge.* London (paginated as pp. 40–75 of *An essay for introducing a portable laboratory . . . by Peter Shaw, M.D. and Francis Hauksbee* [the younger]).

———(1734a). *An enquiry into the contents, virtues and uses, of the Scarborough spaw-waters . . .* London.

———(1734b). *Chemical lectures publickly read at London, in the years 1731 and 1732; and since at Scarborough . . .* London.

Siegfried, R. (1963a). The discovery of potassium and sodium, and the problem of the chemical elements. *Isis 54,* 247–258.

———(1963b). Further Daltonian doubts. *Isis 54,* 480–481.

———(1967). Boscovich and Davy: Some cautionary remarks. *Isis 58,* 236–238.

Select Bibliography

Singer, C.; E. J. Holmyard; A. R. Hall; and T. I. Williams (1954–58). Eds. *A history of technology.* 5 vols. Oxford.

Sivin, N. (1962). William Lewis as a chemist. *Chymia 8,* 63–88.

Smeaton, W. A. (1957). L. B. Guyton de Morveau (1737–1816). A bibliographical study. *Ambix 6,* 18–34.

———(1961). Guyton de Morveau's course of chemistry in the Dijon Academy. *Ambix 9,* 53–69.

———(1962). *Fourcroy. Chemist and revolutionary.* Cambridge.

———(1963a). Guyton de Morveau and chemical affinity. *Ambix 11,* 55–64.

———(1963b). New light on Lavoisier: the research of the last ten years. *Hist. Sci. 2,* 51–69.

———(1964). Guyton de Morveau and the phlogiston theory. In Cohen and Taton (1964), I, 522–540.

———(1967). Louis Bernard Guyton de Morveau, F. R. S. (1737–1816) and his relations with British scientists. *Notes Rec. R. Soc. Lond. 22,* 113–130.

See also Gibbs and Smeaton (1961).

Smith, R. A. (1856). *Memoir of John Dalton, and history of the atomic theory up to his time.* London.

Smith, R. W. I. (1932).* *English-speaking students of medicine at Leyden.* Edinburgh.

Smyth, A. L. (1966). *John Dalton (1766–1844). A bibliography of works by and about him.* Manchester.

Snow, A. J. (1926). *Matter and gravity in Newton's physical philosophy . . .* London.

Spencer, J. B. (1967–68). Boscovich's theory and its relation to Faraday's researches: an analytic approach. *Archs. Hist. exact Sci. 4,* 184–202.

Stahl, G. E. (1723).* *Fundamenta chymiae dogmaticae & experimentalis . . .* Nuremberg.

———(1730). *Philosophical principles of universal chemistry . . . Drawn from the Collegium Jenese of G. E. Stahl.* London. (A translation of Stahl [1723], by P. Shaw.)

Stephen, L. (1876). *History of English thought in the eighteenth century.* 2 vols. London (reprinted, New York, 1962).

Stewart, A. G. (1901). *The academic Gregories.* London.

Stewart, D. (1854–60). *Collected works.* Edited by Sir William Hamilton. 11 vols. Edinburgh.

Strong, E. W. (1951). Newton's "mathematical way." *Jnl. Hist. Ideas 12,* 90–110.

———(1957). Newtonian explication of natural philosophy. *Jnl. Hist. Ideas 18,* 49–83.

Select Bibliography

Taton, R. (1964). Ed. *Enseignement et diffusion des sciences en France au XVIIIe siècle*. Paris.
    See also Cohen and Taton (1964).
Taylor, B. (1712). Part of a letter from Mr. Brook Taylor, F. R. S. to Dr. Hans Sloane R. S. Secr. Concerning the ascent of water between two glass planes. *Phil. Trans. R. Soc.*, no. 336, 538.
———(1715). An account of an experiment made by Dr. Brook Taylor assisted by Mr. Hawkesbee, in order to discover the law of the magnetical attraction. *Phil. Trans. R. Soc.*, no. 344, 294–295.
———(1721). Extract of a letter from Dr. Brook Taylor, F. R. S. to Sir Hans Sloane, dated 25 June, 1714. Giving an account of some experiments relating to magnetism. *Phil. Trans. R. Sec.*, no. 368, 204–208.
Taylor, E. G. R. (1954). *The mathematical practitioners of Tudor and Stuart England. 1485–1714.* Cambridge.
———(1966). *The mathematical practitioners of Hanoverian England, 1714–1840.* Cambridge.
Taylor, F. S. (1956). An alchemical work of Sir Isaac Newton. *Ambix 5,* 59–83.
Texas (1968). The *Annus Mirabilis* of Sir Isaac Newton. Tricentennial Celebration. *Texas Quart. 10,* no. 3.
Thackray, A. W. (1966a). The origins of Dalton's chemical atomic theory: Daltonian doubts resolved. *Isis 57,* 35–55.
———(1966b). Documents relating to the origins of Dalton's chemical atomic theory. *Mem. Proc. Manchester lit. phil. Soc. 108,* no. 2.
———(1966c). The emergence of Dalton's chemical atomic theory: 1801–08. *Br. Jnl. Hist. Sci. 3,* 1–23.
———(1966d). The Newtonian tradition and eighteenth-century chemistry. Cambridge University Ph.D. thesis no. 5848.
———(1967). Fragmentary remains of John Dalton. Part 1: Letters. *Ann. Sci. 22,* 145–174.
———(1968a). Quantified chemistry—the Newtonian dream. In Cardwell (1968), 92–108.
———(1968b). "Matter in a nut-shell": Newton's *Opticks* and eighteenth-century chemistry. *Ambix 15,* 29–53.
———(in press). "The business of experimental philosophy": the early Newtonian group at the Royal Society. (To appear in *Act. XIIe Cong. int. Hist. Sci.*)
Thomson, J. (1832). *An account of the life, lectures and writings of William Cullen, M.D.* Vol. 1. Edinburgh (Vol. 2. Edinburgh, 1859).

Select Bibliography

Thomson, T. (1802). *A system of chemistry.* 4 vols. Edinburgh.
———(1808). On oxalic acid. *Phil. Trans. R. Soc. 98,* 63–95.
———(1825). *An attempt to establish the first principles of chemistry by experiment.* 2 vols. London.
———(1830–31). *A history of chemistry.* 2 vols. London.
Thorpe, T. E. (1902). *Essays in historical chemistry.* London.
Torlais, J. (1937).* *Un Rochelais grand-maître de la francmaçonnerie et physicien au XVIIIe siècle. Le Révérend J. T. Desaguliers.* La Rochelle.
Turnbull, H. W. (1959–61). Ed. *The correspondence of Isaac Newton.* Vols. *1–3, 1661–1694.* Cambridge. (See also Scott [1967].)
Turner, E. (1825). *An introduction to the study of the laws of chemical combination and the atomic theory. Drawn up for the use of students.* Edinburgh.
Turner, G. l'E. (1967). A portrait of James Short, F. R. S., attributable to Benjamin Wilson, F. R. S. *Notes Rec. R. Soc. Lond. 22,* 105–112.
Van Melsen, A. G. (1952). *From atomos to atom. The history of the concept "atom."* Pittsburgh, Pa.
Venel, G. F. (1753). Article "Chimie" in the *Encyclopédie. Vol. III.* Paris.
Villamil, R. de (1931). *Newton—the man.* London.
Voltaire, F. M. A. de (1733a). *Lettres écrites de Londres sur les Anglois et autres sujets.* Amsterdam. (Published 1734.)
———(1733b). *Letters concerning the English nation.* London.
———(1738a). *Elémens de la philosophie de Newton, mis à la portée de tout le monde.* Amsterdam.
———(1738b). *The elements of Sir Isaac Newton's philosophy. By Mr. Voltaire.* Translated from the French. Revised and corrected by John Hanna. London.
Walker, A. (1802). *A system of familiar philosophy . . . including every material modern discovery . . . A new edition.* 2 vols. London.
Wasserman, E. R. (1965). Ed. *Aspects of the eighteenth century.* London.
Watson, R. (1771). *A plan of a course of chemical lectures.* Cambridge.
———(1781–87). *Chemical essays.* 5 vols. London.
Wenzel, C. F. (1782). *Lehre von der Verwandschaft der Körper.* Dresden (originally published 1777).
Westfall, R. S. (1958). *Science and religion in seventeenth-century England.* New Haven, Conn.

Select Bibliography

———(1962). The foundations of Newton's philosophy of nature. *Br. Jnl. Hist. Sci. 1,* 171–182.

———(1966). Newton's optics: the present state of research. *Isis 57,* 102–107.

Wheeler, T. S. See Partington and Wheeler (1960).

Whitehead, J. (1778). *Materialism philosophically examined; or the immateriality of the soul asserted and proved, on philosophical principles; in answer to Dr. Priestley's disquisitions . . .* London.

Whiteside, D. T. (1962). The expanding world of Newtonian research. *Hist. Sci. 1,* 16–29.

———(1967–). Ed. *The mathematical papers of Isaac Newton.* (To be completed in eight volumes.) Cambridge.

Whittaker, E. (1951). *A history of the theories of aether and electricity. Vol. 1. The classical theories.* Revised edition. London.

Whyte, L. L. (1961). Ed. *Roger Joseph Boscovich . . . Studies of his life and work . . .* London.

Wigham, H. M. (1890). A bag of old letters. *Friends Quart. Exam. 24,* 545–558; *25,* 406–416; *26,* 77–92; *27,* 515–527.

Wightman, W. P. D. (1955–56). William Cullen and the teaching of chemistry. *Ann. Sci. 11,* 154–165; *12,* 192–205.

Wilkinson, T. T. (1855). An account of the early mathematical and philosophical writings of the late Dr. Dalton. *Mem. Proc. Manchester lit. phil. Soc. 17,* 1–30.

Willey, B. (1940). *The eighteenth-century background.* London.

Williams, L. P. (1960). Faraday and the structure of matter. *Contemp. Phys. 2,* 93–105.

———(1965). *Michael Faraday. A biography.* New York.

Wilson, B. (1746). *An essay towards an explication of the phaenomena of electricity, deduced from the aether of Sir Isaac Newton . . .* London.

Wilson, J. (1761). Ed. *Mathematical tracts of the late Benjamin Robins Esq.* 2 vols. London.

Wohl, R. (1960). Buffon and his project for a new science. *Isis 51,* 186–199.

Wolf, A. (1952). *A history of science, technology, and philosophy in the eighteenth century.* 2nd ed. London.

Wollaston, W. H. (1808). On super-acid and sub-acid salts. *Phil. Trans. R. Soc. 98,* 96–102.

Wordsworth, C. (1877). *Scholae academicae: some account of the studies at the English universities in the eighteenth century.* Cambridge.

Young, T. (1807). *A course of lectures on natural philosophy and the mechanical arts.* 2 vols. London.

Young, W. (1793). Ed. *Contemplatio philosophica: a posthumous work of the late Brook Taylor. To which is prefixed a life of the author . . . With an appendix containing sundry original papers . . .* London.

Index

Achard, Franz Carl, 212, 220
Acidity, 173, 194
Acids, 94, 98, 99, 168, 200, 239, 242
Acta Eruditorum, 54, 58, 59, 71
Adams, George, 236, 250, 251, 256, 273, 274
Adhesion, 212, 213, 228
Affinity: force of, 5, 94, 97, 121, 159, 201, 206–208, 211–216, 222, 227, 239, 241, 243–244, 257, 261–262, 267, 275; of metals, 37; tables of, 38, 73, 74, 90, 202–204, 215, 217, 219, 223, 229, 266; diagrams, 224, 226–227, 241. *See also* Elective attraction
Aix-la-Chappelle, 78
Algarotti, Francesco di, 83
Algebra in chemistry, 70–72, 218, 224, 241
Alkalinity, 173
Alkalis, 94, 98, 99, 168, 239, 242
Amalgamation of metals, 220
Amsterdam, 87, 93, 222
Ardern, James, 236
Atmosphere: constitution of, 261; pressure of, 143, 258
Atoms: chemical atoms, 7, 13, 15, 23, 26, 39, 57, 64, 105, 109, 131, 233, 248, 249, 251, 264, 264n, 271, 277, 279; physical atoms, 15, 143, 144, 161, 163, 176, 184; size of, 177, 273; weight of, 40, 241, 263, 268

Bacon, Roger, 119
Banks, John, 236, 254
Barner, Jacob, 172
Barometers, 212
Baron, Théodore, 206
Baumé, Antoine, 205, 210, 217
Beattie, James, 191, 248
Becher, Johann Joachim, 172, 201
Beddoes, Thomas, 219, 224
Bentley, Richard, 16
Bentley, Letters to, 30, 32
Bergman, Torbern, 1, 5, 121, 213, 214, 215, *218–221*, 228, 229, 242
Berkeley, George, Bishop of Cloyne, 10, 57, 133, 245, 248
Berlin, 212, 223
Bernoulli, John, 89, 193
Berthollet, Claude Louis, 42, 95, 122, 213, *230–233*, 238, 259, 275
Berzelian system of notation, 266n
Bewley, George, 253, 254
Birch, Thomas, 125, 139
Black, Joseph, ix, 40, 107, 122, 140, 147, 149, 168, 176, 186, 192, 215, *224–226*, 240, 242
Boerhaave, Herman, 9, 46, 101, 104, *106–113*, 116, 119, 120, 121, 122, 140, 149, 150n, 171, 182, 185, 193, 206, 224, 242
Boscovich, Roger Joseph, 10, 26, 31, 32, 68, 69, 125, 131, 132, 134, 142, 145, 146, 149, *150–155*, 160, 188, 190, 191, 214, 249

319

Index

Boyle, Robert, 21, 24, 71, 111, 112, 119, 162, 165, 169, 174, 176, 184, 193, 200
Boyle, Letters to, 26, 30, 31, 112, 125, 139
Boyle's law for gases, 33
Brewster, Sir David, 8
Brunet, Pierre, 84
Bucquet, Jean-Baptiste-Marie, 217
Buffon, George-Louis Leclerc, Comte de, 5, 31, 32, 69, 95, 97, 98, 100, 117, 121, 125, *155–160,* 195, 205, 208, 210, 211, 212, 214, 216, 219, 222, 231, 281
Burgundy parliament, 210
Butterfield, Herbert, viii, 172n

Calcination, 71, 211
Calculus, Leibnizian, 60
Calculus dispute, 49, 52, 54, 88, 156
Caloric, 40, 135, 139, 142, 150. *See also* Imponderable fluids
Cambridge, 45, 46, 77, 114, 126, 224
Campbell, Rev. Colin, 48n, 55n
Campbell, George, 247
Capillary rise, 75, 77, 78, 86, 104, 157, 158. *See also* Surface tension
Caroline, Princess, 61
Cartesian science, ix, 56, 82, 92, 96, 99, 112, 125, 127, 163, 180, 203
Cavendish, Henry, 121, 240
Chambers, Ephraim, 110
Chandos, Duke of, 235
Châtelet, Madame du, 83, 100, 157
Chemical balance, 197, 208
Chemical compounds, 173, 176, 186, 207, 209, 228, 231, 240, 264, 267, 272, 277
Chemical equivalents, 223, 229, 232, 233, 239, 264, 277
Chemical mechanisms, 3, 68, 100, 122, 158, 163, 167, 171, 175, 190, 199, 203, 206–208, 221, 230, 239, 240, 241, 243, 257, 262, 275
Chemical mixtures, 173, 228
Chemical molecules, 239, 264. *See also* Particles
Chemical qualities, 173, 194
Chemical reaction rates, 222, 228
Chemical specificity, 170, 199, 202

Chemical units, 199, 207. *See also* Atoms; Particles
Cheyne, George, 46, 48, *49–51,* 68, 73, 75
Cigna, Giovanni Francesco, 211
Clairaut, Alexis Claude, 98, 121, 156, 157–158, 201, 208, 212, 218
Clarke, Samuel, 61. *See also* Leibniz-Clarke correspondence
Cleghorn, William, *147–150,* 151, 191
Cohen, I. Bernard, 10n, 150
Cohesion, 23, 35, 67, 72, 100, 109, 118, 139, 141, 145, 164, 181, 232, 237, 276
Colden, Cadwallader, 141
Collinson, Peter, 141
Color, *see* Light
Combustion, 147, 169, 173, 198
Commercium Epistolicum, 59
Condillacean philosophy, 217
Corpuscular philosophy, 105, 110, 143, 179, 193, 245
Coste, Pierre, 84, 93
Cotes, Roger, 16, 88, 114, 126–127
Craig, John, 46
Crosland, Maurice, 7n
Cullen, William, 107, 122, 135, 139, 147, 175, 175n, 185, 186, 194–195, 200, 205, 206, 215, 224, 226, 240–241, 242, 281

D'Alembert, Jean Le Rond, 96, 196, 201
Dallowe, Timothy, 110
Dalton, John, 2, 3, 6, 25, 34, 39, 40, 41, 107, 113, 130, 146, 150, 151, 198, 201, 204, 221, 223, 233, 236, 238, 241, 243, 248, 250, 251, *252–278,* 280
Dalton's atomic theory, 204, 248, 256, 261, 264, 277, 279
Darby, Abraham, 253
Daubeny, Charles, 277
Davy, Humphry, 40, 151, 154, 191, 204, 261, 266, 273, 275, 277
Demachy, Jacques-François, 218
Density, *see* Specific gravity
Desaguliers, John Theophilus, 19, 66, 74, 77, 80, 83, 93, 100, 101,

Index

114, 117, 122, 136, 209, 212, 235, 236
Descartes, René, 13, 100, 108, 127, 152, 245
Destructive distillation, 167, 174
Diderot, Denis, 196
Dijon Academy, 210
Directed valencies, 154, 264
Dissenting academies, 16, 235, 256
Double refraction, 183
Dublin, 135, 139, 147, 227
Du Fay, Charles François de Cisternay, 114
Duisberg, 105
Dutch science, 88, 113, 179, 182

Edinburgh, 45, 46, 49, 107, 140, 147, 191, 224, 242, 262
Elasticity, 67; of air, 33, 35, 38, 143; of solids, 35, 145; of water, 56; of fluids, 138, 140
Elective attraction, 202, 207, 216, 219, 221, 224, 241. See also Affinity
Electricity, 31, 32, 36, 40, 73, 75, 112, 122, 125, 135, 139, 147, 149, 150, 237, 247, 277; static electricity, 36, 75, 81; electrochemistry, 272, 275. See also Voltaic pile
Elements: chemical, 42, 110, 116, 131, 151, 161–199 passim, 238, 251n, 271, 273, 275; Empedoclean, 143; physical, 185. See also Principles
Eller, Johann Theodor, 196
Elliott, John, 240, 242
Enlightenment, chemists of the, 1, 41, 191, 221–224, 235, 241
Ether, 26–32, 75, 111, 122, 125, 130, 133, 135–140 passim, 237, 247; particles of, 138
Evaporation, 258

Faraday, Michael, 151, 280
Fatio de Diullier, Nicolas, 43
Fichman, Martin, 193n
Fire, 32, 104, 106, 111, 122, 149, 155, 169, 182, 246. See also Heat
Fischer, Ernst Gottfried, 223
Flamsteed, John, 254

Fontenelle, Bernard Le Bovier de, 20, 84, 88, 89, 90, 93, 97, 109, 202
Force-curves, 103, 152, 155, 215
Fordyce, George, 213, 241, 242
Fothergill, John, 253
Fourcroy, Antoine-François, Comte de, 40, 121, 176, 202, 209, 213, 215, 217, 220, 243
Franklin, Benjamin, 112, 125, 135, 139
Freind, John, 5, 37, 42, 44, 46, 50, 51, 58, 59, 68, 70, 73, 87, 94, 95, 97, 114, 117, 119, 121, 126, 155, 163, 175, 192, 193, 208, 209, 212, 215, 242
French Académie Royale des Sciences, 74, 78, 85, 90, 96, 156; *Mémoires*, 96, 158; *Histoire*, 97
French Revolution, 231, 250, 281
French science, 49, 53, 74, 83, 84, 88, 92, 95, 150n, 157, 158, 176, 189, 192, 196, 201, 215, 216, 241, 250, 280
Fusion of metals, 81

Garnett, Thomas, 122, 236
Gassendi, Pierre, 13
Gentleman's Diary, 255–256
Geoffroy, Etienne-François, 37, 74, 85, 89, 90, 92, 121, 202, 204, 206, 224
Geometric progressions, 223
Geometry, 216
George I, King of England, 101
German science, 53, 55, 84, 88, 172, 222, 223, 241, 281
Gillispie, Charles Coulston, 194n
Glasgow, 140, 191, 205
Gough, John, 255
'sGravesande, William Jacob Storm van, 66, 83, 100, *101–104*, 113, 118, 136, 209, 254
Gravity, 14, 16, 23, 26, 29, 31, 36, 38, 67, 68, 81, 109, 130, 139, 141, 142, 145, 150, 157, 159, 181, 201, 203, 208, 209, 212, 220, 247
Gray, Stephen, 114
Green, Robert, 17, *126–134*, 151, 246
Gregory, David, 27, 34, 43, 44, *45–48*, 55, 62, 68

321

Index

Guerlac, Henry, 4n, 196n
Guyton de Morveau, Baron Louis-Bernard, 1, 2, 80, 95, 197, 210, 214, 216, 219, 220, 222, 226, 228, 229, 276

Hales, Stephen, 34, 40, 100, *114–118*, 122, 128, 156, 175, 193
Halle, 172
Halley, Edmond, 177, 254
Hamilton, William, 151
Hanover, 62n. See also Sophia, Electress of Hanover
Harris, John, 24, 44, 54, 58, 59, 122, 135, 165, 234
Hartsoeker, Nicolaus, 59
Harvey, William, 137
Hauksbee, Francis, 28, 37, 58, *74–80*, 86, 95, 104, 105, 114, 116, 214, 235, 237
Hauksbee, Francis (the younger), 235. See also p. 312 above under Shaw (1731)
Heat, 32, 35, 40, 147–150, 171, 221, 232, 258, 268, 275; conduction of, 81
Heat capacity, 149
Helmont, Franciscus Mercurius van, 196
Helsham, Richard, 136
Henry, William, 107, 185n, 262
Herschel, William, 151
Hickman, Nathaniel, 156
Higgins, Bryan, 122, 221, 227, 241, 243, 277
Higgins, William, 221, 241, 242, 243, 277
Hippocrates, 200
Hobbes, Thomas, 126, 245
Hodgson, James, 235
Hoefer, Ferdinand, 2n
Hoffmann, Friedrich, 172
Holland, 83, 84, 101, 113. See also Dutch science
Homberg, Wilhelm, 119, 167, *169–171*, 200, 240
Hooke, Robert, 75
Horne, George, Bishop of Norwich, 146
Horsley, John, 236
Hume, David, 188, 245, 248

Hutchinson, John, 246
Hutchinsonian group, 17, 131, 146, 247, 280. See also Adams; Horne; Hutchinson; Jones

Imponderable fluids, 9, 30, 31, 40, 112, 125, *134–150*, 246, 251. See also Electricity; Ether; Fire; Phlogiston
Industrial Revolution, 107–108, 204, 235, 236. See also Manufactures; Technological developments
Inertia, 15
Inertial mass, 69. See also Matter
Inflammability, 194
Internal structure of matter, 2, 9, *12–26*, 39, 58, 63–66, 99, 138, 140, 143, 149, 164, 167, 172, 175, 177, 187, 190, 249, 274; porosity, 5, 18, 21, 22, 35, 53, 55, 57, 63–65, 109, 112, 144, 177, 178, 181, 189, 208, 247; vacuum, 16, 18, 22, 56, 57, 59, 61, 127, 129, 177; plenum, 62, 66, 127, 129. See also "Nut-shell" theory and Diagram I, p. 19 above
International context of science, 4, 69, 89, 101, 280
Inverse power laws, 31, 38, 77, 97, 98, 155, 157–160, 205, 212, 213
Italy, 83

Jardin du Roi, 205, 211
Jena, 171
Jones, William, 246, 251, 274
Journal des Savants, 86, 87, 93, 94
Journal de Trévoux, 86, 87, 93, 94, 102

Kant, Immanuel, 222
Keill, James, 42, 46, 51, 57, 58, 68, 114, 116, 121
Keill, John, 42, 44, 48, 51, 52, 53, 57, 58, 59, 60, 65, 67, 71, 73, 77, 89, 94–96, 114, 120, 121, 154, 155, 189, 192, 193, 208, 209, 212, 215, 218, 245
Kepler, Johannes, 277
Kirwan, Richard, 2, 5, 175, 215, 219, 220, 222, 223, 224, 226
Knight, David, 168n

Index

Knight, Gowin, 26, 139, *141–147*, 148, 151, 152
Kopp, Hermann, 2, 2n

Ladies' Diary, 253, 256
Lagrange, Joseph-Louis, Comte de, 92, 98, 201, 212
Laplace, Pierre-Simon, Comte de, 31, 92, 98, 121, 196, 197, 201, 217, 232
Latent heat, 149
Lavoisier, Antoine-Laurent, viii, ix, 2, 4, 6, 34, 39, 41, 92, 117, 150, 168, 175n, 176, 188, 196, 198, 201, 210, 211, 215–217, 230, 258, 272, 280; correspondence of, 4n
Laws of motion, 49
Leclerc, George-Louis, see Buffon
Leibniz, Gottfried Wilhelm, Baron von, 13, 28, 44, 49, 52–62 *passim*, 81, 82, 88, 89, 151
Leibniz-Clarke correspondence, 52, 61, 65
Lemery, Nicholas, 24, 68, 72, 94, 126, 165–169 *passim*, 200, 206, 225
Lewis, William, 242
Leyden, 46, 50, 100, 105, 106, 140, 156
Liebig, Justus von, 278
Light: nature of, 18, 32, 40, 93, 100, 122, 147, 150, 155, 169, 170, 178, 182, 246; reflection of, 14, 20, 36, 38, 129; refraction of, 14, 36, 129; transmission of, 14, 22, 53, 63, 122, 178, 190; color, 14, 20, 74, 81, 173, 176, 186, 194; color and particle size, 20–21, 23
Loadstone, 78, 81
Locke, John, 126, 127, 133, 245
Lower, Richard, 137

Mach, Ernst, 8
Macquer, Pierre Joseph, 1, 95, 176, 186, 195, 197, 204, *205–210*, 212, 216, 219, 231, 239
Magellan, J. J., 229
Magnetism, 21, 32, 36, 40, 66, 78, 109, 141, 143, 147, 150, 277
Magnus, Albertus, 199–200
Manchester, 252, 256, 259, 267, 271

Manchester Literary and Philosophical Society, 259, 261, 263, 266
Manuel, Frank, 8n
Manufactures, relationship of scientific theory to, ix, 1, 48, 50, 107, 184, 195, 204, 205, 276, 281. *See also* Industrial Revolution; Professionalization of science; Technological developments
Mariotte, Edme, 86
Martin, Benjamin, 122, 135, 178, 179, 210, 236, 237
Mathematics, relationship between sciences and, 2n, 5, 38, 44, 48, 201, 204, 218, 222, 223, 234, 252, 258, 270
Matter: inertial homogeneity of, 5, 16–18, 22, 122, 124, 134, 152, 161, 251, 251n, 269–270; indestructibility of, 29, 110, 163, 171, 272, 273; transmutation, 14, 22, 24, 29, 175n, 187, 196, 198; inertial heterogeneity of, 39–40; infinite divisibility of, 59, 65, 102, 143, 251; indivisibility of, 99, 152, 173, 272
Maupertuis, Pierre Louis Moreau de, 96–99 *passim*, 121, 156, 157, 193, 201, 208, 209, 254
Mead, Richard, 46, 48, *50–51*, 68, 73, 106, 113, 119, 120, 121
Medicine and chemistry, 48, 49, 50, 68, 69, 70, 106–109, 136–138, 141, 147, 149, 204, 205
Meissen porcelain factory, 222
Metallicity, 173, 181–182, 203
Metallurgy, 204
Meteorology, 253, 255, 257
Michell, John, 151
Mickleburgh, John, 114
Mixed gases, theory of, 258, 259, 262, 267, 268, 275
Monmort, Remond, Comte de, 74, 84, 85, 89, 90
More, Henry, 13, 60n
Moyes, Henry, 236, 237
Musschenbroek, Peter van, 83, 100, *105–106*, 113, 181, 209, 254

Napoleon Bonaparte, 231
Nervous impulse, 147, 150
Neumann, Caspar, 172

323

Index

Newton, Isaac, 1–4, 8–42, 45, 49, 52, 57, 58, 69, 71, 74, 76, 80, 83, 85, 88, 89, 105, 110, 114, 115, 116, 119, 124, 135, 136, 146, 157, 177, 183, 184, 189, 193, 208, 212, 242, 274, 277, 280; textual changes in manuscripts by, 11, 15, 17, 22, 27, 31, 45, 48, 59, 75, 82, 86, 124; and the Royal Society, see Newtonian circle; Royal Society; use of Queries in *Opticks*, 14, 18–20, 22–23, 24, 27, 28, 30, 33, 35–38, 48, 55, 69, 75, 93, 116, 151
Newtonian circle, 41, 42, 55, 56, 57n, 65, 82, 120, 126, 281
Newtonian tradition, 3, 5–7, 15, 31, 34, 40, 41, 83–84, 121–123, 151, 160, 161, 203, 214, 250, 258, 276–279; in Britain, 40, 48, 113–120, 175, 188, 205, 225, 235, 242, 255, 266, 269, 274; in France, 4, 84–100, 117, 176, 208, 210, 211, 216, 232; in Holland, 101–113; in Sweden, 214, 219; political overtones, 88, 89, 101. *See also* Theology: influence on science
Nicholson, William, 240, 243, 272
Nollet, Jean-Antoine, abbé, 125
"Nut-shell" theory, 3, 22, 41, 53–67, 82, 143, 176, 178, 188, 245, 281

Occam's razor, 131
Odorizers, 176
Oldenburg, Letters to, 30
Olson, Richard, 191n
Optical instruments, 183
Organic chemistry, 267, 275, 278
Orleans, Duke of, 169
Oxford, 45, 48, 54, 70, 80, 141, 224

Paris, 78, 93, 99, 105, 156, 189, 210
Particles, 33, 200, 237; shape of, 15, 69, 97, 103, 142, 144, 151, 159, 163, 170, 181, 186, 200, 209, 214, 237, 264; size of, 20, 170, 177, 178, 186, 200, 209, 237, 273; of first composition, etc., 23–24, 64, 142, 154, 164, 181, 190; ultimate, 25, 144, 184; spheres of activity of, 103, *see also* Force-curves; primary, 142, 144; weight of, 263, *see also* Relative weights. *See also* Atoms; Internal structure of matter
Partington, James Riddick, 2, 172n
Pearson, George, 242
Pemberton, Henry, 66, 84, 114, 120, 254
Philosophical Transactions, 57, 59, 60, 76, 77, 78
Phlogiston, 4, 40, 111, 119, 139, 143, 149, 150, 168, 172, 173, 186, 229, 230, 244
Pitcairne, Archibald, 45, 48, 49, 50, 55, 106
Plummer, Andrew, 224
Pneumatic chemistry, 168, 196, 231
Political context, 85, 89, 101, 121, 198, 231, 246–247, 250–252, 280–281
Political controversy, 3, 44, 52–53, 55, 62n, 88–89, 188, 280–281
Popular lecturers, 6, 44, 54, 80, 107, 120, 204, 234–238, 250, 252, 254, 267, 281
Powers, 56, 189, 209, 231. *See also* Affinity; Short-range forces
Price, Richard, 249
Priestley, Joseph, 3, 10, 31, 53n, 54, 134, 146, 151, 175, 188, 189, 197, 215, 247, 248, 250, 280, 281
Primary qualities, 16, 30, 32
Principles: chemical, 116, 163, 165, 166, 169, 170, 173, 175–176, 186, 197–198, 207, 273; physical, 174. *See also* Elements
Professionalization of science, viii, ix, 48, 50, 83, 107, 204–205, 222

Quakers, 6, 141, 252, 253n, 281

Rain, 257
Reductionist philosophy, vii, 6, 38, 44, 49, 71, 83, 163, 170, 175, 176, 179, 186, 193, 196, 197, 200, 203, 211, 217, 279
Reid, Thomas, 191, 248
Relative weights, 263, 266, 276
Reynolds, Richard, 253
Richter, Jeremias, 121, 221, 222, 228, 230

Index

Robinson, Bryan, 30, 32, 125, *135–141*, 147, 186
Robinson, Elihu, 253
Robison, John, 151, 191, 225
Romanticism, ix
Rotheram, Caleb, 236
Rouelle, Guillaume-François, 193
Rouelle, Hilaire-Martin, 193, 197, 211
Rouen, Academy of, 206
Rowning, John, 83, 103, 136, 145
Royal Institution, 267, 271, 273
Royal Society, 28, 33, 36, 42, 44, 49, 93, 95, 96, 111; president of, 44, 74, 276; elections, 49, 51, 77, 80, 85, 89, 96, 101, 115, 156, 227; *Journal Book,* 52, 57n, 66, 77, 79; secretary of, 58n, 78, 89, 234; curator of experiments, 66, 74, 78, 80, 235; other, 44, 74, 80, 81, 85, 113, 116, 121, 141, 227, 229, 235, 275, 276
Rozier, abbé François, 211
Rutherforth, Thomas, 83, 254
Ryswick, peace of, 85

Scheele, Carl Wilhelm, ix
Schofield, Robert E., 3, 149n
Secondary qualities, 16, 130
Senac, Jean-Baptiste, 94n
Shaw, Peter, 110, 114, 116, *118–120*, 121, 171, 184, 185, 206, 236, 281
Short-range forces: attraction, 24, 28, 31, 51, 55–58 *passim*, 72, 85, 86, 90, 94, 96, 98, 103, 118, 122, 132, 143, 145, 148, 151, 170, 186, 202, 203, 211, 213, 216, 218, 221, 226, 231, 247, 257; repulsion, 28, 31, 103, 118, 122, 132, 143, 145, 148, 151, 186, 211, 259, 262, 264, 279; other, 5, 9, 15, 16, 32–38, 40, 41, 42, 50, 67–73, 74, 75, 79, 81, 87, 95, 99, 100, 140, 160, 162, 163, 201, 207, 209, 232, 237, 258, 269
Sloane, Hans, 59, 74, 90
Smeaton, William, 211
Socinian tradition, 134
Solubility, 35, 72, 155, 256–258, 261, 262, 268
Sophia, Electress of Hanover, 52
Spanish Succession, War of, 85, 89

Specific gravity, 17, 71–72, 128, 181, 208, 227, 228, 251, 273
Specific heat, 149
Spinoza, Benedictus de, 126, 245
Stahl, Georg Ernst, 119, 143, 168, 171, 176, 184, 185, 192, 197
Stahlian theory, ix, 3, 174n, 195, 196, 201, 203, 206, 217, 280. *See also* Phlogiston
Stewart, Dugald, 151, 191, 248
Stukeley, William, 114
Sublimation, 72
Succession to English throne, 52, 55
Surface tension, 38, 79, 157, 158, 213
Sweden, 214, 241
Swiss science, 88

Taylor, Brook, 74, 77, 80, 81, 85, 89, 95, 105, 114, 211, 214, 215
Technological developments, 80, 235, 281
Teddington, 115
Theology: influence on science, 6, 10, 31, 49, 61, 101, 131, 134, 138, 142, 145, 171, 188, 234, 244, 246, 247, 248, 251, 252, 255, 257, 280; influence on Newton, 13, 15, 16, 23, 27, 30, 56
Thomson, Thomas, 2, 151, 154, 192, 238, 261, 267, 276, 281
Turin, 211
Turner, Matthew, 236

Uppsala, 214, 218
Utrecht, 50, 105

Vacuum, *see* Internal structure of matter
Varignon, Pierre, 84, 88n, 93
Venel, Gabriel-François, 176, 186, 193, 194, 197, 200, 203, 211, 214, 215
Vienna, 151
Voltaic pile, 191, 275. *See also* Electricity
Voltaire, François-Marie Arouet, 83, 96, 99, 104, 156, 179, 208, 209
Von Meyer, Ernst, 2n
Vortex mechanisms, 84

Walker, Adam, 122, 126, 256n

Index

Waller, John, 114
Warltire, James, 236, 237
Warwick, Rev. Dr. Thomas, 237
Water vapor, 257; pressure of, 258
Watson, Richard, 224
Wedel, G. W., 172
Wenzel, Carl Friedrich, 215, 221, 228, 230
Wesley, John, 146, 248
Whewell, William, 266n
Whiston, William, 44, 114, 235
Whitehead, John, 146, 249
Wilson, Benjamin, 135, 139, 141
Wolff, Christian, 58, 60
Wollaston, William Hyde, 268, 276
Wordsworth, William, 255
Worster, Benjamin, 235
Wren, Sir Christopher, 26